Environmental Consulting Fundamentals

Investigation and Remediation

Environmental Consulting Fundamentals

Investigation and Remediation

Benjamin Alter

CRC Press
Taylor & Francis Group
Boca Raton   London   New York

CRC Press is an imprint of the
Taylor & Francis Group, an **informa** business

CRC Press
Taylor & Francis Group
6000 Broken Sound Parkway NW, Suite 300
Boca Raton, FL 33487-2742

© 2012 by Taylor & Francis Group, LLC
CRC Press is an imprint of Taylor & Francis Group, an Informa business

No claim to original U.S. Government works
Version Date: 20120501

International Standard Book Number-13: 978-1-4398-6891-1 (eBook - PDF)

Visit the Taylor & Francis Web site at
http://www.taylorandfrancis.com

and the CRC Press Web site at
http://www.crcpress.com

In memory of my father,

who passed away during the preparation of this book.

He was more than a father—he was a father-and-a-half.

Contents

Section II   Site Investigations and Remediations

Section III   Land Usage

Section IV Indoor Environmental Concerns

Preface

This book was born of necessity.

In 2000, I began teaching a graduate course, entitled "Environmental Investigations and Remediations" at Hunter College in New York City. I designed the course to be a survey course that would provide someone interested in becoming an environmental consultant upon graduation with the basic building blocks needed to start a career in this dynamic, multidisciplinary field. There was only one problem—I could not find a textbook to support the course. There were textbooks that covered only one-third of the course material, but none that covered the other two-thirds of the course. I was forced to cover that material with a combination of regulations and guidance documents that were, in general, far too detailed and obtuse for a survey course, or with material written for "citizens" that was far too simplistic.

So, in the course of the 10 years that I taught the course, I started to fill in the reading assignments with my own writings, which led to me pursuing the "fully Monty," namely this book.

This is the book that I wish was available when I was designing the course. In fact, it is the book I wish someone had written when I began my career as an environmental consultant! It is designed for the student as well as the beginning practitioner.

Each chapter covers a topic that merits a book itself; however, they are not intended to make the reader an expert in any of the topics presented. Nor, for that matter, will the reader be able to go out and perform an environmental investigation or remediation after reading this book. That skill requires not just the knowledge obtained from reading this book, but the field experience, ideally obtained under the apprenticeship of a more experienced hand.

Rather than placing an emphasis on formulas, equations, and regulatory requirements, this book emphasizes the *thought processes* that go into designing an environmental study, interpreting the data obtained from the study, and selecting the next step, be it further investigation or remediation. It also discusses the specific roles played by the environmental consultant, and the roles played by others in the investigation and remediation activities, thereby giving the reader the "big picture" of how these activities actually happen in real life.

The book begins with an overview of the environmental consultant—the typical consultant's educational background, formal training, and on-the-job training. It follows with brief summaries on three of the important building blocks of environmental investigations and remediations—their regulatory structure in the United States, and the scientific underpinnings of the processes that occur in the environment. The next several chapters take the reader through the steps of subsurface investigations and remediations, going from Phase I and Phase II environmental site assessments through to remediation. This sequence of chapters is followed by a chapter on ecological studies and a chapter on environmental impact assessments, a huge subfield of environmental consulting.

The next set of chapters takes the reader indoors, as they cover environmental issues related to buildings—being from the Northeast and teaching in the fall semester, this transition dovetailed nicely with the ensuing seasonal change! The topics covered include asbestos, lead-based paint, radon, mold, and indoor air quality. The book concludes with a short chapter on describing a typical environmental consulting project, including designing the scope of work and developing a prospective budget and project schedule.

Connecting many of the chapters are examples of environmental problems at a fictitious factory and office building. The examples are designed to put flesh on some of the theoretical concepts presented in the book. They also provide a thread through which many of the book's chapters can become connected in the mind of the reader.

Because of its diversity and ever-changing landscape, I have never stopped learning the varied aspects of environmental consulting. Neither will you. Enjoy the journey! I am pleased you have chosen this book as one of your starting points.

Acknowledgments

Numerous people assisted me in the preparation of this book. I would like to thank Dr. Paul Davis, Dr. Larry Feldman, Stephen Lecco, Dr. Chun-Hua Liu, and Ben Sallemi of GZA GeoEnvironmental, Inc. for their critical review of various portions of the book. I would also like to thank Dr. Howard Apsan of City University of New York; Dr. Stephen Konigsberg of Adventus Americas; Professor Richard Mendelsohn of Rutgers University; Professor Monica Tischler of Benedictine University; Will Moody of Geo-Cleanse International, Inc.; and Carrie Anne Vinch for their reviews of portions of the book. . I'd like to especially acknowledge Marilyn Rose, www.marilynrosedesign.com, a talented artist who did the artwork for this book. The photograph on the front cover was taken by Frank Vetere, my colleague at GZA and an excellent photographer in his spare time. GZA GeoEnvironmental, Inc. gets a big thank you for its support and encouragement of my project. Last but not least, I would like to thank my wife, Jean, for always believing in me.

The Author

Benjamin Alter has been an environmental consultant since 1989. Prior to becoming a consultant, he was a geophysicist in the oil industry, tasked with exploring for oil and natural gas along the Gulf Coast. He has a Bachelor of Science degree from the State University of New York at Albany, a Master of Science degree from Cornell University, and an MBA from Columbia University.

As an environmental consultant, Alter has designed and managed multiple remedial investigations and remediations, and has conducted hundreds of site assessments throughout the United States. He has provided litigation services for numerous hazardous waste cases, and has been a key contributor in the development of New Jersey's Licensed Site Remediation Professional (LSRP) program. He was an adjunct professor at the Hunter College School of Health Sciences in New York City from 2000 to 2009, and has published numerous articles and given numerous presentations on environmental investigations and remediations. Alter is currently a principal and Senior Vice President with GZA GeoEnvironmental, Inc. in its Fairfield, New Jersey office.

Section I

Environmental Consulting
A Perspective

1

What Is Environmental Consulting?

1.1 The Environment and Environmental Hazards

To understand what constitutes environmental consulting, we first must understand the meaning of "the environment." *Webster's Dictionary* defines *environment* as

> the complex of physical, chemical, and biotic factors (as climate, soil, and living things) that act upon an organism or an ecological community and ultimately determine its form and survival.

Let's dissect this definition and discuss how it pertains to the contents of this book.

As the definition indicates, physical factors include climate and soil, where climate includes the air, sunlight, and one of the fundamental requirements for life on earth (and a topic of discussion in many of the book's chapters), water. The chemical factors include the interactions between many of these physical factors as well as chemicals that occur naturally and those introduced by mankind. The "living things" indicated in the definition encompass the full range of living things: microbial, plant, and animal life.

Conditions that have the ability to affect these living things are known as *environmental hazards*. An environmental hazard should not be confused with chemicals that can adversely change the environment. These chemicals, known in various contexts as *pollutants* or *contaminants*, are one of the three essential parts of an environmental hazard. For an environmental hazard to exist, three conditions must be present (see Figure 1.1). There must be a *source* of the pollution, a *receptor* for the pollution, and a *pathway* connecting the two.

Although it might appear convenient to place receptors into one of the three categories of living things (microorganisms, plants, and animals), it is better to place receptors into one of two categories: humans and everything else. There is a profound emphasis on humans as potential receptors in the environmental regulations (discussed in Chapter 2). In fact, many of the

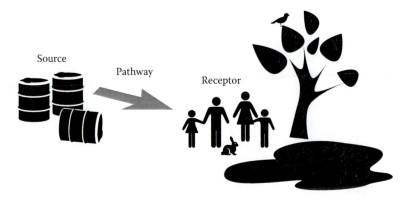

FIGURE 1.1
The three components of an environmental hazard.

environmental hazards discussed in this book (such as asbestos and lead-based paint) are hazards only to humans and no other living beings.

Sources of contamination can be quite varied. They may be natural sources, such as rock formations that contain lead, arsenic, or other naturally occurring toxins. They may be chemicals or petroleum products that were manufactured for the benefit of humans but are toxic to humans and often a wide variety of biota. They may be building materials such as the above-mentioned asbestos and lead-based paint that were manufactured for the benefit of humans but are toxic to humans exposed to these materials.

Within the category of receptors, there is a subset of receptors, known as *sensitive receptors*, which are so named because of their heightened ability to be impacted by contamination. Sensitive human receptors typically include children, the elderly, and people with disabilities, and properties where these populations typically are found, such as schools, parks, and nursing homes. Depending on the contaminant, this category might also include immuno-compromised people, people with allergies, and other such populations. In the environment, sensitive receptors may include wetlands (discussed in Chapter 9), surface water bodies, rain forests, endangered and threatened species, and their habitats.

The most common contaminant pathways leading to receptors in the environment include:

Water—Water can transport contaminants from the source as solutes (chemicals dissolved in water) or as particulates, which are solids suspended in water. In general, water is considered either "surface water" (e.g., rivers, lakes, and saltwater bodies), and "groundwater," which is water entrained in geologic formation. Groundwater is further discussed in Chapter 4.

Soil and rock—Contaminants can move through the void spaces of soil and rock. See Chapter 4 for an analysis of this pathway.

Overland pathways—Chemical spills, especially in liquid form, can migrate on the surface through overland paths, especially if the surface is paved or otherwise impermeable to percolation into the underlying earth.

Air—Air can transport contaminants over great distances. Contaminants and particulates released from volcanic eruptions have the ability to cross the globe and circle it several times. Air pollution can arise from both stationary sources, such as power plants, and mobile sources, such as motor vehicles.

When one considers the hundreds of thousands if not millions of chemicals, both naturally occurring and man-made, that exist in the world today, the billions of humans and the countless number of other biota, and the various pathways to connect the two groups, one can start to imagine the infinite amount of permutations that can constitute an environmental hazard.

1.2 What Is Environmental Consulting?

To define the second part of the term environmental consulting, we go back to *Webster*. The *Webster* dictionary defines *consulting* as "providing professional or expert advice." This advice is provided to people and companies, who, in this context, are known as *clients* (types of clients are described later in this chapter). So we hereby define *environmental consulting* as "providing professional or expert advice regarding the environment to clients."

In practice, however, it is a lot more than that. In its broadest sense, the role of the environmental consultant is to minimize or eliminate, or assist in the minimization or elimination of, environmental hazards, or demonstrate that they do not exist. That covers a lot of ground.

Because there are an infinite amount of possible environmental problems, the skill set that environmental consultants bring to the table vary widely. In this book, the term "environmental consultant" is used almost interchangeably with the terms "consultant," "geologist," "chemist," "biologist," "engineer," and "investigator." That is because an environmental consultant may be any one of these things or all of these things depending upon the situation and the type of project. They also may be known by the particular role they are playing on a project, such as "air monitor," "field supervisor," or "risk assessor." The role of the environmental consultant can even be more complicated than this, as discussed below.

1.2.1 Environmental Consulting Services

One basic type of service provided by environmental consultants involves inspections, assessments, and studies to identify areas where the client is not in compliance with existing regulations (Chapter 2 provides a summary of this regulatory framework). Because of the wide range of activities performed by environmental consultants, providing a comprehensive study of these services could be a book in itself. Some of the broader categories of services provided by environmental consultants are discussed in this and the subsequent section.

The United States has a body of environmental regulations that is complex and often confusing. In printed form, it is many thousands of pages in length and supplemented by guidance documents, databases, and spreadsheets and formulae that dwarf the regulations upon which they are based. Much of the environmental consultant's job is to guide the client through this labyrinth.

Often, assisting the client in obtaining compliance begins with an *environmental compliance* audit, in which an environmental consultant studies a client's operations to assess whether they are in compliance with environmental regulations. This assessment may be comprehensive or targeted to one or more environmental issues. It may target one facility, a portion of one facility, multiple facilities, or the entire company operations. Alternatively, it may not target facilities at all, but rather environmental policies, practices, and procedures established at the corporate or government bureau level or maybe the division level. Once problems (or opportunities) are identified, the consultant will proceed with assisting the client in designing systems, preparing regulatory-required plans or documents, or training workers to attain regulatory compliance.

Environmental consultants support the vast infrastructure of environmental services required by clients throughout the world. Many of these entities are involved in the management of wastes, including the handling, transportation, and processing of wastes. These clients must comply with various environmental regulations, which often require the services of consultants who specialize in providing these services. Consultants assist these facilities in their design and construction; their initial permitting and subsequent permitting; and ultimately the design and implementation of their closure. Many of the larger waste-handling firms have in-house experts who assist them in achieving and maintaining regulatory compliance. However, many times these "in-house" experts are actually environmental consultants retained by private sector companies on a full- or part-time basis (known as "outsourced employees") to supply their expertise where needed.

Environmental consultants may perform studies to determine the potential effects of a planned action on the environment. This may include performing some of the analyses required in the preparation of *environmental impact statements* (EIS), which is discussed in Chapter 10.

In recent years, private sector companies have sought to go beyond the regulations in their environmental "performance." Voluntary governmental initiatives in waste minimization and recycling have led to the formation of a "green" economy, which includes various entities that want to minimize their impact on the environment. These entities may be interested in being perceived as green for marketing purposes, to save money, to minimize the liability associated with the generation of wastes, or simply to be good global citizens.

An international initiative started in 1997 by the International Standards Organization, or ISO, encouraged companies to design and implement *environmental management systems*, which are management systems designed to minimize impacts on the environment. The guidelines developed by ISO, known as the *ISO 14000 series*, have been adopted by thousands of corporations throughout the industrialized world. Environmental consultants assist private entities in formulating and implementing these voluntary measures, and auditing the success of these measures. Consultants may also assist corporations in *benchmarking*, a method by which the company's environmental performance is compared to other companies in the same industry, companies of similar size, or companies from the same geographical area.

Environmental consultants who, due to their education, training, or experience, have developed expertise and specialized knowledge in technical or regulatory matters, may provide support to attorneys who are involved in litigation regarding an environmental matter. Such consultants, known as *expert witnesses*, assist attorneys in litigating civil, and occasionally criminal, lawsuits. Consultants will produce expert reports in support of either the party that alleges damage (the plaintiff) or the party that was allegedly responsible for the damage (the defendant). Part of the consultant's litigation support services is providing a *deposition*, which is sworn testimony given outside of the courtroom to attorneys on both sides of the litigation. Environmental consultants may also assist the attorney in preparing the various pieces of written and verbal correspondence that take place in the course of the litigation, may take part in calculating the assessment of monetary damages, and may take the stand in a trial.

Emerging sectors in the field of environmental consulting include sustainability, green building design and construction, green product claims, climate change, and renewable energy. As with other voluntary environmental initiatives, environmental consultants assist clients in these endeavors in the design of implementation plans, implementing operational changes at a facility or company-wide, auditing the effectiveness of these changes, and benchmarking the client's performance.

1.2.2 The Consultant as Contractor

A large portion of the work and certainly the lion's share of the revenue earned by environmental consultants deals with the impact of pollution on

Training for environmental consultants is quite varied as well. In general, there is no set level of training required of environmental consultants, although there are regulatory requirements that depend upon the type of consulting services to be provided. For instance, consultants who will perform fieldwork relating to hazardous waste remediation are required by federal law to take a 40-hour course known as Hazardous Waste Operations and Emergency Response (HAZWOPR). There are a wide variety of licenses required for consultants working in the fields of asbestos and lead-based paint as well as licenses dealing with radon investigation and mitigation.

Environmental consultants often have certifications that allow them to perform certain functions on a project. For instance, the Professional Engineer (PE) license, granted by states or territories, is often required for environmental consultants responsible for the design of remediation systems and other structures. In most cases, the signature of a PE is required on a system design before a regulatory entity will allow its construction. A Professional Geologist (PG) certification is granted by some states and many independent associations, such as the American Institute of Professional Geologists (AIPG).

A consultant holding a Certified Hazardous Materials Manager (CHMM) license, which is issued by the Institute of Hazardous Materials Management (IHMM), certifies individuals with expertise in the management of hazardous materials. A Certified Industrial Hygienist (CIH) license, which is issued by the American Board of Industrial Hygiene (ABIH), is given to individuals with expertise in worker safety. Since worker safety is closely intertwined with environmental investigations and remediations, CIHs play many roles in such projects.

Certain states issue licenses to qualified environmental consultants, and require investigations and cleanups of hazardous wastes and petroleum to be conducted, overseen, and in some cases certified by the licensed individual. Among the states that issue such licenses are California (Registered Environmental Assessor), Massachusetts (Licensed Site Professional [LSP]), Connecticut (Licensed Environmental Professional [LEP]), and most recently New Jersey (Licensed Site Remediation Professional [LSRP]). As is the case with PE, PG, and other professional licenses, the individual must first qualify for these licenses based on educational background and years of relevant experience in the field, and usually must pass a comprehensive test to obtain the license.

Continuing education (CE) is a necessary aspect of environmental consulting. Some training courses require annual or biennial updates. For instance, consultants with HAZWOPR training are required to take an 8-hour refresher course annually, as are consultants with asbestos inspector licenses and lead paint risk assessor licenses. Environmental consultants holding one of the aforementioned licenses typically have continuing education requirements associated with their licenses. In most cases, they are required to take several courses relevant to their certification in a given time period so that they can maintain their licenses. These courses are taught by academics, such as professors as well as environmental consultants with the requisite expertise.

The next chapter lays out the regulatory framework with which clients in the United States must comply.

2

Framework of Environmental Regulations

2.1 The "Nature" of Environmental Regulations

Perhaps more than almost any other business, the business of environmental consulting is driven by regulations. A complete description of all federal environmental regulations would require its own book. In this chapter, we limit our discussion to the legal framework, history, and scope of the major federal environmental regulations that pertain to the topics discussed in subsequent chapters of this book.

2.1.1 Prehistory of Environmental Regulations

Prior to 1969, environmental laws were geared primarily to the protection of navigable waterways and the conservation of natural resources. The Rivers and Harbors Act of 1899, the oldest federal environmental law in the United States, prohibited the dumping of refuse into a navigable water body or its tributaries without a permit. This act also prohibited the excavation, filling, or altering of any port, harbor, or channel without a permit.

Beginning with the tenure of Gifford Pinchot in 1898 and the formation of the U.S. Forest Service under the Roosevelt Administration in 1905, the conservation of natural resources became the primary environment-related goal of the federal government. A series of laws aimed toward the conservation of natural resources was passed in the ensuing decades, the most notable of which was the Wilderness Act of 1964, whose mission was to preserve designated national forest lands in their natural condition.

The first federal law aimed at chemical hazards was the *Federal Insecticide, Fungicide, and Rodenticide Act*, known as FIFRA, which was first enacted in 1947 and revised and expanded in subsequent years. FIFRA originally was oriented toward consumer protection and required pesticide manufacturers to register their products. The Delaney Clause of FIFRA, which took effect in 1958, required that manufacturers of consumer products demonstrate that their products would not cause cancer in people or animals.

The first half of the 20th century saw an explosion of growth in the invention and usage of synthetic chemicals in the Western world. Sanitary engineers who were used to dealing with compounds that occur naturally and biodegrade naturally were unaccustomed to dealing with synthetic chemicals that did not biodegrade naturally and therefore persisted in the environment. Many of these chemicals proved to be hazardous to a wide variety of fauna and flora, and posed a growing health threat in the world's industrialized countries.

The general public gained awareness of this threat when Rachel Carson published *Silent Spring* in 1962. In that landmark book, Carson documented the degradation of the environment caused by these chemicals and warned that if we humans are to live among these chemicals we should be aware of their effects on us. Her book helped to launch the modern environmental movement.

2.1.2 Establishment of the USEPA and OSHA

The passage of the *National Environmental Policy Act* (NEPA) in 1969 set the stage for modern environmental regulations. NEPA created the *Council on Environmental Quality* (CEQ), part of the executive branch of government, which acknowledged the "profound impact" of human activity on the environment and the "critical importance of restoring and maintaining environmental quality to the overall welfare and development of man." Chief among the new requirements under NEPA was the requirement to prepare an environmental impact statement (EIS) for actions contemplated by the federal government that may have a "significant impact" on the environmental quality in the area. NEPA is discussed in greater detail in Chapter 10.

Following on the heels of NEPA was the establishment of the *U.S. Environmental Protection Agency* (USEPA) and the *Occupational Safety and Health Administration*, better known as OSHA. The USEPA's mission was to develop a regulatory framework to protect the environment, while OSHA's primary responsibility was to protect the health and safety of workers in the workplace. The responsibilities of these two agencies overlap with regard to workers who investigate and remediate contaminated properties, and this book discusses OSHA regulations within the context of environmental investigations and remediations.

Charged with protecting the environment in the United States, the USEPA set to the task to develop a regulatory framework to protect the environment. In its first six years of existence, the USEPA passed six major laws, with major amendments and reauthorizations following in subsequent years. Table 2.1 is a partial list of major federal environmental laws passed since 1970. With one exception, all of these major laws are forward looking, that is, designed to prevent future environmental degradation. The one backward-looking law is the Comprehensive Environmental Response, Compensation, and Liability Act, more commonly known as Superfund. These major laws are discussed below.

TABLE 2.1

Major Environmental Laws after the Establishment of the U.S. Environmental
Protection Agency

1970	Clean Air Act Amendments
1972	Federal Water Pollution Control Act Amendments (Clean Water Act)
	Federal Environmental Pesticides Control Act of 1972 (amended FIFRA)
1973	Endangered Species Act
1974	Safe Drinking Water Act
1976	Resource Conservation and Recovery Act (RCRA)
	Toxic Substances Control Act (TSCA)
1977	Clean Air Act Amendments
	Clean Water Act Amendments
1980	Comprehensive Environmental Response, Compensation, and Liability Act (Superfund)
1984	Hazardous and Solid Waste Amendments (RCRA Amendments)
1986	Safe Drinking Water Act Amendments
	Superfund Amendments and Reauthorization Act (also known as Emergency Planning and Right-To-Know Act [EPCRA])
1987	Water Quality Act
1990	Clean Air Act Amendments of 1990
	Oil Pollution Act of 1990
	Pollution Prevention Act
2003	Brownfields Act
2005	Energy Policy Act

Source: Adapted from Kraft, Michael E., 1996, *Environmental Policy and Politics*, HarperCollins College Publishers.

2.2 Major Federal Environmental Laws

2.2.1 Clean Air Act

The *Clean Air Act*, originally passed in 1963, actually predates the USEPA. However, the original law provided federal support for air pollution research and had little impact on private or public sector entities. The Clean Air Act Amendments of 1970 vastly strengthened the Clean Air Act and included enforcement provisions, providing some of the regulatory structure that exists today.

The 1970 amendments established *National Ambient Air Quality Standards* (NAAQS) for certain *criteria pollutants*. These pollutants included sulfur dioxide (SO_2), nitrogen oxides (NO_x), which includes nitrous oxide (NO_2) and nitric oxide (NO_3), ozone (O_3), carbon monoxide (CO), and PM_{10}, which is an abbreviation for particulate matter with a diameter of 10 microns or less (this threshold for particulates was set as a practical limit of attainment rather than a scientific limit based on health risk).

The 1977 Clean Air Act Amendments added to the attainment requirements in nonattainment areas by calling for the prevention of significant deterioration (PSD) for attainment areas. These amendments also established three classes of "clean areas":

Class I—National parks and similar areas, where air quality is to be protected from any deterioration

Class II—A specified amount of deterioration would be permitted

Class III—Air pollution could continue to deteriorate, but not beyond national standards

In addition, a standard for lead was added in the Clean Air Act in 1977.

The Clean Air Act Amendments of 1990 introduced sweeping changes to the Clean Air Act. These amendments included:

1. The passage of the National Emissions Standards for *hazardous air pollutants* (HAPs), known as *NESHAPs*, which increased the number of hazardous air pollutants with NAAQS to 187 (including asbestos; see Chapter 12), plus acid rain and chlorofluorocarbons (CFCs). The USEPA identified nonattainment areas for HAPs and required the states to devise plans to meet criteria for nonattainment areas in their respective jurisdictions.

2. The establishment of emissions standards for mobile and stationary sources.

3. A requirement that new air pollution sources, such as new power plants and new car models, must use best available technology (BAT) to comply with the standards set for the criteria pollutants.

4. The establishment of an emissions trading program, through which the owners of pollution sources in nonattainment areas could trade pollution "credits" with owners of pollution sources in attainment areas, thereby meeting their regulatory requirements.

5. Requirements for oxygenated fuel, which would result in more efficient fuel combustion and therefore less air pollution during the cold winter months. In an interesting case of the solution to one problem leading to the creation of a different problem, this requirement gave rise to the widespread use of methyl tertiary butyl ether (MTBE) (see Chapter 6).

6. The establishment of the *Title V permitting program*, which applies to major stationary sources of air pollutants. A major stationary source is defined by its air emission rates and depends upon whether the source is in an attainment area, a nonattainment area, or an area of serious nonattainment for the specific HAP in question. A Title V permit describes the air emissions, calculates the theoretical air emis-

sions possible from the facility, specifies the type and performance of existing air emissions control equipment, and includes an air monitoring plan.

Environmental consultants who specialize in air permitting are often involved in the preparation, renewal, and updating of Title V air permits as well as the design of air pollution control systems for regulated air emissions sources. They also measure and quantify air pollutants for the purposes of compliance with the Clean Air Act, establishing the effectiveness of air emissions control technologies, as well as establishing credits to be traded in the emissions trading program.

2.2.2 Clean Water Act

The *Clean Water Act*, passed in 1972 and amended in 1977 and 1987, regulates the discharge of pollutants into the navigable waters of the United States. At first, the Clean Water Act focused on *conventional pollutants*, so named because of the decades-long focus on these pollutants by sanitary engineers. "Conventional" pollutants are discussed in Chapter 9.

The 1977 amendments to the Clean Water Act forced the USEPA to focus on toxic pollutants as well as conventional pollutants. In response, the USEPA created a list of *priority pollutants* to be regulated under the Clean Water Act (see Chapter 3). Dischargers of conventional and priority pollutants to surface waters were required to obtain a permit from the USEPA under the *National Pollutant Discharge Elimination System* (NPDES). Each NPDES permit defines the *total maximum daily load* (TMDL) allowed for each toxic pollutant. Regular monitoring and self-reporting of wastewater discharges to navigable water bodies is required under a NPDES permit. The 1977 amendments to the Clean Water Act also provided funding for municipal wastewater treatment plants.

The 1987 amendments to the Clean Water Act (known as the Water Quality Act) added so-called *nonpoint sources* of water pollution to the discharge regulations. Such nonpoint sources include *storm water runoff*, which lacks a specific "point source" on a site, such as a discharge pipe, but nonetheless can add pollutants to a water body by sweeping up contaminants and debris lying on the surface and carrying them into the nearby water body. Companies with the ability to create nonpoint source pollution were required to develop Storm Water Pollution Prevention Plans (SWPPP) as a best management practice.

Section 404 of the Clean Water Act covers the dredging and filling of *wetlands*, which are discussed in Chapter 9.

The Clean Water Act provides work opportunities for environmental consultants in assisting facilities in obtaining NPDES permits and monitoring wastewater discharges for compliance with the Clean Water Act.

Consultants also prepare SWPPPs and assist in the identification and mitigation of wetlands.

2.2.3 Endangered Species Act

The *Endangered Species Act* (ESA) is administered by the U.S. Fish and Wildlife Service (USFWS) as well as the National Oceanic and Atmospheric Administration (NOAA). The ESA defines a species as endangered if it is "in danger of extinction throughout all or a significant portion of its range." It defines a species as threatened if it is "likely to become an endangered species within the foreseeable future."

The ESA's primary goal is to prevent the extinction of imperiled plant and animal life; its secondary goal is to recover and maintain those populations by removing or lessening threats to their survival. Under the ESA, endangered animals are protected even on private land, although plants are not. When a proposed action has the ability to impact an endangered species, the ESA requires studies to assess and quantify the potential impacts, and implementation of measures to minimize or eliminate those impacts. A study of endangered species is required for all EISs (see Chapter 10), and has been known to result in the delay or even cancellation of major construction projects whose implementation would be detrimental to endangered and threatened species.

2.2.4 Safe Drinking Water Act

The *Safe Drinking Water Act* (SDWA), first passed in 1974 and amended in 1986, regulates some 50,000 permanent water systems throughout the United States. Only permanent water systems that have at least 15 services connections or service at least 25 people are regulated under the SDWA to avoid an unnecessary regulatory burden on small businesses and homeowners with private wells. The SDWA sets national drinking water standards for a host of chemical, biological, physical, and radiological agents (see Chapter 3 for further discussion of these standards), including lead (see Chapter 14). The SDWA requires the system operator to utilize best available technology to achieve drinking water standards and establishes secondary drinking water standards for publically owned treatment works (POTWs), such as sewage treatment plants.

2.2.5 Toxic Substances Control Act

The *Toxic Substances Control Act* (TSCA), passed by Congress in 1976, requires the testing of new chemicals as well as chemicals already in commerce to assess their effect on human health and the environment. Of special note was the TSCA ban of the manufacture of polychlorinated biphenyls (PCBs) in 1979. Other chemicals that are regulated under TSCA (as well as other statutes) include asbestos, beryllium, hexavalent chromium, dioxins, and dibenzofurans. Lead-based paint, discussed in Chapter 13, became regulated under TSCA in the Residential Lead-Based Paint Hazard Reduction Act of 1992.

Title II of TSCA, which became law in 1986, is better known as the Asbestos Hazard Emergency Response Act, or AHERA. This law regulates asbestos-containing building materials in schools (see Chapter 11). Title III of TSCA regulates the abatement of elevated radon concentrations in indoor air (see Chapter 15).

Environmental consultants assist regulated entities in compliance with TSCA regulations, including the investigation and remediation of PCBs, asbestos, and other TSCA-regulated substances. They also assist in the testing and verification of new chemicals in accordance with TSCA regulations.

2.2.6 Resource Conservation and Recovery Act

The *Resource Conservation and Recovery Act* (RCRA), which Congress passed in 1976, is a deceptively named environmental law. Although its name implies a relatively benign mission such as plastic and paper recycling, its scope goes far beyond that.

The objectives of RCRA broadly fall into two categories: (1) preventing hazardous wastes from entering the environment; and (2) minimizing the generation of all types of solid wastes, which is defined as a waste that can be deposited in a landfill, including gaseous, liquid, and semiliquid waste. RCRA governs the storage, transportation, and disposal of all solid wastes.

2.2.6.1 Definition of Hazardous Waste

The term "hazardous waste" was first defined under RCRA, and later used by other laws, especially the *Comprehensive Environmental Response, Compensation, and Liability Act* (CERCLA) (see Section 2.2.7). Its definition under RCRA is described next. Under RCRA, hazardous waste can be either *listed hazardous waste* or *characteristic hazardous waste*. Listed hazardous wastes, as implied by the name, appear on one of several lists developed by the USEPA. These wastes are considered hazardous *per se*, regardless of their physical or chemical properties. Under RCRA, the following wastes are designated as hazardous wastes:

- Process wastes from general industrial processes (the *F list*)
- Process wastes from specific industrial processes (the *K list*)
- Unused or off-specification chemicals, container residues and spill cleanup residues of acute hazardous waste chemicals (the *P list* and the *U list*)

- Other chemicals, a category that includes a broad spectrum of waste chemicals deemed to be hazardous. These wastes, known as *universal wastes*, include typical residential and commercial wastes such as batteries, pesticides, mercury-containing equipment, and bulbs from electric lamps.

Characteristic hazardous wastes (*D list* wastes) are not listed wastes but have physical characteristics that are considered to be hazardous to human health. For a nonlisted waste to be considered hazardous, it must have at least one of the following four characteristics:

Ignitable (flammable)—Flash point below 140°F (60°C). This means that the material can vaporize and spontaneously catch on fire without the benefit of a spark or flame. Materials with flash points above 140°F (60°C) that will catch on fire when exposed to a spark or flame are labeled as combustible. D001 is the code given to ignitable waste.

Corrosive—pH below 2 or above 12.5. D002 is the code given to corrosive waste.

Reactive—Normally unstable, reacts violently with air or water, or forms potentially explosive mixtures with water; emits toxic fumes when mixed with water or is capable of detonation. D003 is the code given to reactive waste.

Toxic—A waste that tests toxic using the *toxicity characteristic leachate procedure* (TCLP). TCLP is a laboratory procedure that tests the ability of a chemical to leach out of the solid matrix in which it is present and enter the environment.

The tests that are required under RCRA to test for hazardous waste characteristics include: flash point (to address the waste's ignitability), pH, and the waste's reactivity. In addition, RCRA requires analysis for certain volatile organic compounds (VOCs), semivolatile organic compounds (SVOCs), pesticides, and metals (these chemical classes are defined in Chapter 3). The 40 RCRA compounds/analytes are shown in Table 2.2. Wastes that test hazardous are given codes D004–D043, one code for each contaminant involved.*

RCRA exempts certain types of waste from hazardous waste characterization, such as household waste (other than those mentioned earlier), agricultural waste returned to the ground, and utility wastes from coal combustion.

---

* The USEPA assigned hazardous waste codes D004–D043 in alphabetical order rather than the order shown on Table 2.2.

TABLE 2.2

Compounds Required to Be Tested to Determine Whether a Solid Waste Exhibits the Characteristic of Toxicity

VOCs	SVOCs	Pesticides	Metals
Benzene	Nitrobenzene	Chlordane	Arsenic
Chlorobenzene	Hexachlorobenzene	2,4-D	Barium
1,4-Dichlorobenzene	Hexachloroethane	Endrin	Cadmium
1,2-Dichloroethane	Hexachlorobutadiene	Heptachlor (and its epoxide)	Chromium
Tetrachloroethylene	Pyridine	Lindane	Lead
Trichloroethylene	m-Cresol	Methoxychlor	Mercury
1,1-Dichloroethylene	o-Cresol	Toxaphene	Selenium
Vinyl chloride	p-Cresol	2,4,5-TP (Silvex)	Silver
Methyl ethyl ketone	Cresol		
Chloroform	2,4-Dinitrotoluene		
Carbon tetrachloride	Pentachlorophenol		
	2,4,5-Trichlorophenol		
	2,4,6-Trichlorophenol		

Source: U.S. Environmental Protection Agency.

2.2.6.2 *"Cradle-to-Grave" Concept of Hazardous Waste Management*

RCRA seeks to prevent hazardous wastes from entering the environment at the point of generation, at the point of final disposition, and all points in between. There is a litany of regulations, known as *Subtitle C*, that govern hazardous waste generators and receivers. Subtitle C codifies the most famous of RCRA concepts, namely the "Cradle-to-Grave" concept of hazardous waste management. Under the Cradle-to-Grave concept, the generator of hazardous waste is legally responsible for the waste from the time of generation and forevermore. Therefore, once the wastes are removed from their facility of origin, they remain the responsibility of the owner. It also requires every party that handles or stores hazardous waste to obtain an operating permit from the USEPA, so that proper handling and storage of the waste can be tracked and enforced.

Regulations governing *large quantity generators* (LQGs), which generate more than 1000 kg of hazardous waste per month, dictate how the wastes are stored at their point of origin, where they can be stored, the characteristics of the waste storage area, labeling of the containers, how long they may be stored at their point of origin, and many other aspects of the management of this waste. Less complex regulations apply to *small quantity generators* (SQGs), which generate between 100 kg and 1000 kg of hazardous waste per month, and *conditionally exempt SQGs*, which generate less than 100 kg of hazardous waste per month.

A *Uniform Hazardous Waste Manifest* must accompany each shipment of hazardous waste from its point of generation to its point of disposal. Figure 2.1

shows an example of a hazardous waste manifest. The manifest has a unique number in the right-hand corner for quick identification. It contains information about the generator, the transporter(s), and the designated receiving facility for the waste. It also contains information about the waste, including waste code and quantity.

Each time the waste changes hands, the manifest is signed by a representative of the entity taking over the responsibility for the material. Under RCRA, each of these entities has a legal responsibility for its involvement in the hazardous waste. This part of the process was designed to eliminate "midnight dumping," in which irresponsible, unlicensed parties would accept waste and then dump it at an unauthorized, uncontrolled location rather than incur the expense of bringing it to a properly licensed receiving facility.

Subtitle C also establishes the qualifications for receiving facilities, known in the parlance of RCRA as *treatment, storage, and disposal facilities* (TSDFs). There are elaborate and detailed regulations governing the construction and management of hazardous waste landfills and other TSDFs. The process by which a TSDF obtains its operating license is known as a *Part B* permit.

2.2.6.3 Nonhazardous Waste Management

Nonhazardous solid wastes are regulated under *Subtitle D* of RCRA. Subtitle D is analogous to Subtitle C, in that it regulates the generation, transportation, and disposal of nonhazardous solid waste. Subtitle D also contains elaborate and detailed regulations governing the construction and management of landfills and other receiving facilities.

In 1984, RCRA Subtitle I was passed as part of the RCRA Hazardous and Solid Waste Amendments. Subtitle I established regulations for underground storage tanks (USTs) that store "regulated substances," including hazardous chemicals and petroleum products. Many USTs are exempted from RCRA, including residential heating oil tanks with a capacity less than 1100 gallons, septic tanks, and fuel tanks on farms (some state regulations are more stringent than the federal regulations).

2.2.6.4 RCRA and the Environmental Consultant

RCRA generates a variety of opportunities for environmental consultants, including facility compliance with RCRA hazardous waste regulations; permitting and operating of Subtitle C and Subtitle D facilities, especially the management of Part B permits for Subtitle C facilities; and the many facets of RCRA that directly or indirectly influence the investigation and remediation of contaminated properties. Environmental consultants also investigate and remediate RCRA Corrective Action sites (*CORRACTS*), which are TSDFs that have had contaminant releases to the subsurface.

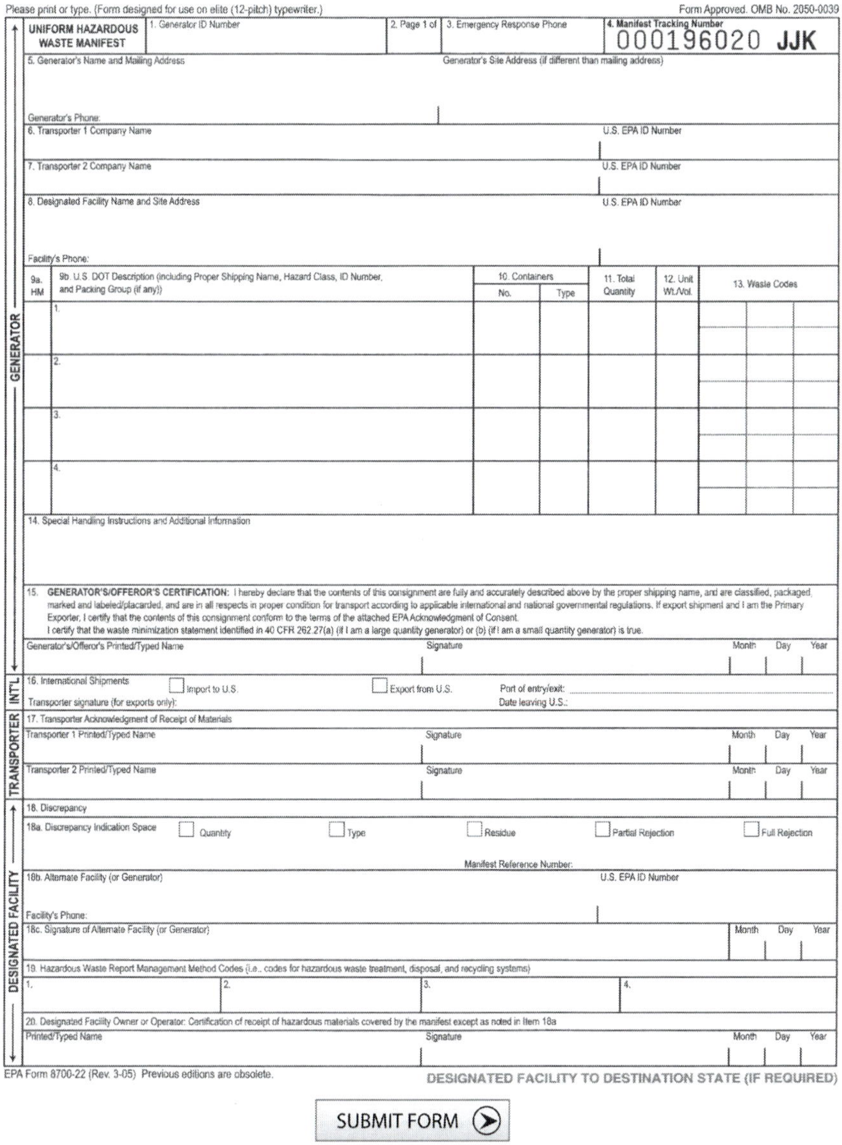

Please print or type. (Form designed for use on elite (12-pitch) typewriter.)

Form Approved. OMB No. 2050-0039

UNIFORM HAZARDOUS WASTE MANIFEST

1. Generator ID Number
2. Page 1 of
3. Emergency Response Phone
4. Manifest Tracking Number 000196020 JJK

5. Generator's Name and Mailing Address

Generator's Site Address (if different than mailing address)

Generator's Phone:

6. Transporter 1 Company Name — U.S. EPA ID Number

7. Transporter 2 Company Name — U.S. EPA ID Number

8. Designated Facility Name and Site Address — U.S. EPA ID Number

Facility's Phone:

9a. HM	9b. U.S. DOT Description (including Proper Shipping Name, Hazard Class, ID Number, and Packing Group (if any))	10. Containers No.	10. Containers Type	11. Total Quantity	12. Unit Wt./Vol.	13. Waste Codes
	1.					
	2.					
	3.					
	4.					

14. Special Handling Instructions and Additional Information

15. **GENERATOR'S/OFFEROR'S CERTIFICATION:** I hereby declare that the contents of this consignment are fully and accurately described above by the proper shipping name, and are classified, packaged, marked and labeled/placarded, and are in all respects in proper condition for transport according to applicable international and national governmental regulations. If export shipment and I am the Primary Exporter, I certify that the contents of this consignment conform to the terms of the attached EPA Acknowledgment of Consent.
I certify that the waste minimization statement identified in 40 CFR 262.27(a) (if I am a large quantity generator) or (b) (if I am a small quantity generator) is true.

Generator's/Offeror's Printed/Typed Name — Signature — Month Day Year

GENERATOR

INT'L

16. International Shipments ☐ Import to U.S. ☐ Export from U.S. Port of entry/exit:
Transporter signature (for exports only): Date leaving U.S.:

TRANSPORTER

17. Transporter Acknowledgment of Receipt of Materials
Transporter 1 Printed/Typed Name — Signature — Month Day Year
Transporter 2 Printed/Typed Name — Signature — Month Day Year

DESIGNATED FACILITY

18. Discrepancy
18a. Discrepancy Indication Space ☐ Quantity ☐ Type ☐ Residue ☐ Partial Rejection ☐ Full Rejection
Manifest Reference Number:
18b. Alternate Facility (or Generator) — U.S. EPA ID Number
Facility's Phone:
18c. Signature of Alternate Facility (or Generator) — Month Day Year

19. Hazardous Waste Report Management Method Codes (i.e., codes for hazardous waste treatment, disposal, and recycling systems)
1. 2. 3. 4.

20. Designated Facility Owner or Operator: Certification of receipt of hazardous materials covered by the manifest except as noted in Item 18a
Printed/Typed Name — Signature — Month Day Year

EPA Form 8700-22 (Rev. 3-05) Previous editions are obsolete.

DESIGNATED FACILITY TO DESTINATION STATE (IF REQUIRED)

SUBMIT FORM

FIGURE 2.1
A blank Hazardous Waste Manifest. (From the U.S. Environmental Protection Agency.)

2.2.7 Comprehensive Environmental Response, Compensation, and Liability Act

The *Comprehensive Environmental Response, Compensation, and Liability Act of 1980*, known officially by its acronym of *CERCLA* but more popularly as *Superfund*, establishes the procedures and standards used in subsurface investigations and remediations.

2.2.7.1 Origins of Superfund

The impetus for the passage of CERCLA in 1980 was a series of major contamination cases that made the front pages in the late 1970s. Thousands of drums that contained hazardous wastes were discovered in what became known as Valley of the Drums in Kentucky. Times Beach, Missouri, made the headlines due to widespread dioxin contamination. However, no hazardous waste site was more infamous than the case known as Love Canal.

Love Canal was a neighborhood in Niagara Falls, New York, that in 1953 was owned by Hooker Chemical. Love Canal was at one time an actual canal constructed by a man named William T. Love as a means to bring hydroelectric power to the area from the falls. His plan failed, and the canal eventually filled with water. In the 1920s, the canal became a dump site for the City of Niagara Falls and during World War II a dump site for the U.S. Army. By the mid-1940s, Hooker Electrochemical Company, later known as Hooker Chemical Company, began dumping chemical waste into the canal.

After World War II, the City of Niagara Falls began expanding in the direction of Love Canal. It asked Hooker to sell the land so that it could build a school on the property. Hooker refused, citing its concerns regarding locating a school over a toxic dump. Hooker eventually yielded to political pressure, selling the land to the city for $1, provided the following appeared in the contract deed (Grantee refers to the City of Niagara Falls, and Grantor refers to Hooker Chemical):

> The Grantee has been advised by the Grantor that the (Love Canal property) has been filled … with waste products resulting from the manufacturing of chemicals … and the Grantee assumes all risk and liability incident to the use thereof … as a part of the consideration of this Conveyance and as a condition thereof, no claim, suit, action, or demand of any nature whatsoever shall ever be made by the Grantee … against the Grantor … in connection with or by reason of the presence of said industrial wastes.

Despite the dire and explicit warning in the contract deed, a neighborhood and a school were built in and around the chemical dump. By the late 1970s, the buried chemicals were identified as the cause of the sickness and death that afflicted the residents of the Love Canal neighborhood. Despite the legal protections seemingly provided by warning in the property deed, widespread publicity and community activism pressured the federal government to take action against Hooker Chemical.

Since no law was available at that time under which action could be brought against Hooker Chemical, Congress responded to public pressure by passing such a law—CERCLA—in 1980. Superfund gave the president of the United States the right to take "any actions necessary to abate … imminent and substantial danger to the public health or welfare or the environment." Superfund delegates to the USEPA the authority to identify parties responsible for the dumping of chemicals at inactive hazardous waste sites and force them to remediate the sites.

2.2.7.2 Liability under Superfund

There are many aspects of Superfund that make it unique among federal laws. The first such aspect of Superfund is *retroactive liability*. Retroactive liability enables Superfund to reach back in time to snare the party responsible for the hazardous waste release with no statute of limitations. Retroactive liability is coupled with *strict liability*, which means that the party responsible for the chemical dumping is liable *even though its actions were permissible when they occurred*. Strict liability is consistent with RCRA's Cradle-to-Grave concept, in that the generator is held liable for the hazardous waste regardless of what party is in possession of the waste. It was the concept of strict liability that enabled USEPA to force Occidental Petroleum, which purchased Hooker Chemical in 1968, to pay for the cleanup at Love Canal.

Last, the Superfund Act created *joint and several liability* for the responsible parties. This means that if there were multiple parties that contributed to an inactive hazardous waste site, the USEPA had the ability to recover all of the costs from any one of the parties, leaving it to that party to recover costs from the other responsible parties.

2.2.7.3 Petroleum Exclusion

The definition of "hazardous substance" under CERCLA specifically excludes petroleum. This portion of the definition, known as the *petroleum exclusion*, appears to exclude contamination caused by any type of petroleum product or crude oil. However, the USEPA interprets this clause as excluding only crude oil and fractions of crude oil, including any hazardous substances, such as benzene, that may be part of the crude oil. Any chemicals added to a crude oil in the refining or manufacturing substance is therefore regulated under CERCLA.

2.2.7.4 National Priorities List

In 1981, the USEPA established the *Hazard Ranking System* (HRS) by which the hazards posed by sites could be assessed to determine whether they should

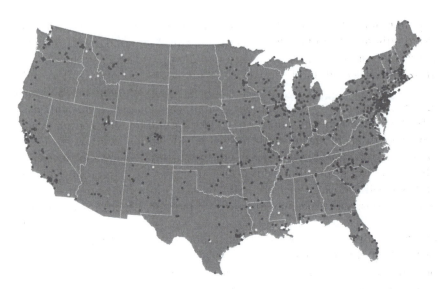

FIGURE 2.2
Superfund sites in the 48 contiguous states as of March 31, 2010. The dark dots are NPL sites and the light dots are proposed NPL sites. (Courtesy of Locus Technologies.)

be included in the Superfund program. This led to the establishment one year later of the *National Priorities List* (NPL), which listed all the sites slated for assessment and remediation under CERCLA. Sites under consideration for inclusion on the NPL were placed on the CERCLA Information Systems list, which is known by its acronym: *CERCLIS*. As of 2010, there were more than 1200 sites on the NPL (see Figure 2.2).

Partial funding for the Superfund program comes from a tax on chemical companies. However, most of the funding comes from USEPA cost recovery actions against *potentially responsible parties* (PRPs) under the "polluter pays principle." USEPA will typically pursue identifiable, available, and financially viable PRPs; these PRPs will generally seek to recoup at least some of their costs by identifying, and sometimes suing, other PRPs. The original law, which was modified in 2003 (see Section 2.2.7.7), contained no provision for *de minimis* contributors, so in theory a pizza parlor whose delivery box was found in a Superfund landfill could be on the hook for the entire cost of the cleanup. These costs typically range on the order of tens of millions of dollars, so determining cost allocation was and is often a costly and bitter process.

2.2.7.5 National Contingency Plan

Activities under CERCLA are governed by the *National Contingency Plan* (NCP), the body of regulations that establishes the framework—and in many cases specific requirements—for subsurface investigations and remediations.

The NCP, or variations on the NCP, have since been used at Superfund and non-Superfund sites. If a site is placed on the NPL, it undergoes a *preliminary assessment/site investigation*, which is analogous to a Phase I/Phase II environmental site assessment (see Chapters 5 and 6), followed by a *remedial investigation/feasibility study* (RI/FS), which is covered in Chapter 7.

The end of the investigation process is the issuance of a *record of decision* (ROD), in which USEPA selects a final remedy. As specified in the 1986 Superfund amendments, this remedy must take into consideration protection of human health and the environment as well as costs and schedule. The statute placed a bias on permanent rather than temporary solutions. Onsite treatment of hazardous wastes was preferred and alternative treatment technologies were encouraged, giving rise to the *Superfund Innovative Technology Evaluation* (SITE) program (see http://www.epa.gov/nrmrl/lrpcd/site/).

2.2.7.6 SARA of 1986

The 1986 amendments to the Superfund Law, known as both the *Superfund Amendments and Reauthorization Act* (SARA) and the *Emergency Planning and Community Right-to-Know Act* (EPCRA), had a dual purpose. SARA also instituted major changes into Superfund. SARA attempted to inject cost considerations into the Superfund proceedings, since costs rose at an alarming rate in the first several years of the program. The act also defined an *innocent purchaser defense*, by which a purchaser of a contaminated property could avoid CERCLA liability. Purchasers who performed "all appropriate inquiries" prior to purchasing the property could now qualify for an innocent purchaser defense, although it did not define the steps needed to obtain this defense.

The other objective of SARA, embodied in Title III of the SARA statute, was a public notification law that was precipitated by the catastrophic chemical release incident in Bhopal, India, in 1984, which killed over 2000 people. EPCRA required industrial facilities to report information about the chemicals they manufactured, stored, and released into the air, which became known as the *toxic release inventory* (TRI).

Also from the Community Right-to-Know portion of SARA came the establishment of the *Local Emergency Planning Committee* (LEPC), which is a group comprised of first responders, such as fire fighters, police, and medical personnel, who might be exposed to hazardous chemicals when responding to an emergency, and would have a need to understand the nature of a person's chemical exposure in order to provide adequate treatment.

A major part of the information provided to the LEPC is the *material safety data* (MSD) *sheet*. MSD sheets are prepared by the manufacturer, and contain toxicity and safety information on the chemicals used at a facility. Any facility that stores or uses hazardous materials must have MSD sheets on hand for anybody who wants to understand what chemicals are present at the facility.

2.2.7.7 Brownfields Act of 2003

The Small Business Liability Relief and Brownfields Revitalization Act of 2003 addressed some of the shortcomings of the original Superfund law. Title I established a *de minimis threshold* for small business contributors, specified exemptions for municipal solid waste (MSW), and provided for expedited settlements with PRPs based on their ability to pay. Title II provided federal grant money for the cleanup of so-called *brownfield* sites, which were defined as urban properties whose development or redevelopment were inhibited by the stigma of actual or perceived contamination. (The redevelopment of brownfield sites rather than "greenfield"—previously undeveloped—sites is being pursued at the federal level as well as by many states and local communities as a strategy for slowing the rate of urbanization.) The Brownfields law also provided a definition for "all appropriate inquiries," which had gone undefined for 17 years (see Chapter 5).

2.2.7.8 Environmental Consulting under CERCLA

Work at Superfund sites employed thousands of consultants nationwide. Consultants are also directly involved in Superfund by providing technical bases for cost allocations. Indirectly, the Phase I industry was created because of Superfund, and Superfund has had a profound influence on the investigation and remediation of all types of contaminated sites, Superfund or not.

2.2.8 Energy Policy Act of 2005

The Energy Policy Act contained a host of provisions relating to energy exploration, consumption, and conservation in the United States. Among those provisions relating directly to the environment are the following:

- It provides incentives for the development of nuclear energy and other energy sources associated with minimal or no emissions of Greenhouse Gases (GHG).
- It encourages the usage of biofuels such as ethanol.
- It includes funding for "clean coal" initiatives, that is, technologies to control the amount of pollutants emitted during the combustion of coal.

The act also prohibited the manufacture and importation of mercury-vapor lamp ballasts after January 1, 2008.

2.3 Legal Framework of Environmental Regulations

2.3.1 Legal Framework of Federal Environmental Regulations

In the United States, laws are passed by Congress and signed by the president. The president also can issue executive orders, which bypass congressional approval. Once a law or an executive order is in place, it is sent to the appropriate department, which is charged with creating regulations to ensure that the directives in the law or executive order are carried out. Once the regulations have been finalized, the department that issued the regulations has the responsibility to enforce them.

At the federal level in the United States, the USEPA is tasked with promulgating and enforcing regulations designed to protect the environment. The USEPA is not a cabinet department, although the USEPA chief administrator is usually given cabinet rank.

Regulations promulgated by the USEPA appear in the *Code of Federal Regulations* (CFR), which can be accessed at www.gpoaccess.gov/cfr/index. html. Each regulation has a numerical title, which refers to the department that administers the regulations. Regulations administered by the USEPA appear in Title 40 of the *Federal Register*. Specific parts of a regulation are ordered by part, section, and subpart. For instance, 40 CFR 761.30(p) is immediately recognizable as an environmental regulation by virtue of its title number. The part of the regulation is 761, the section is 30, and the subpart is (p). OSHA regulations appear in 29 CFR.

The USEPA is divided into 10 regions, as shown in Figure 2.3. Each region is responsible for implementing federal environmental regulations within its jurisdiction. They do not have the authority to issue environmental regulations; that authority resides at USEPA headquarters in Washington, DC.

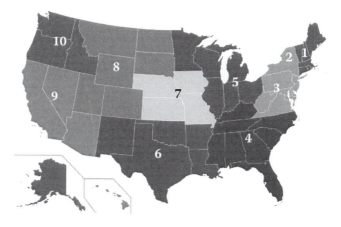

FIGURE 2.3
The 10 regions of the USEPA. (From the U.S. Environmental Protection Agency.)

2.3.2 Legal Framework of State Environmental Regulations

Many of the federal regulations require states to develop plans, known as *state implementation plans,* for the implementation of the federal regulations. Once the USEPA approves a state implementation plan, it is the responsibility of the state to implement the plan, with the USEPA region simply monitoring the success of the state in its implementation of the federal regulation.

Under state implementation plans, the USEPA delegates the enforcement of certain regulations to the states. In such cases, the state environmental protection agencies are responsible for enforcing the USEPA regulations. Appendix B provides a listing of the departments that regulate the environment within their jurisdictions.

Most states promulgate and implement regulations in a manner similar to the USEPA. However, no state can promulgate regulations that are less stringent than the existing federal regulation. When providing environmental consulting services, it is necessary to check on the state and, where applicable, local regulations to ensure full compliance.

States also can pass their own environmental statutes independently from the USEPA. From these statutes, the state environmental regulatory entity typically issues environmental regulations, environmental guidance, and so forth.

References

Bregman, Jacob I. 1999. *Environmental Impact Statements,* 2nd ed. Boca Raton, FL: Lewis Publishers.

Carson, Rachel L. 1962. *Silent Spring.* New York: Houghton Mifflin.

Kraft, Michael E. 1996. *Environmental Policy and Politics.* New York: HarperCollins College Publishers.

Newton, Lisa H., and Dillingham, Catherine K. 1993. *Watersheds: Classic Cases in Environmental Ethics.* Independence, KY: Wadsworth.

U.S. Environmental Protection Agency. December 2010. National Priority List Site Totals by Status and Milestone. www.epa.gov/superfund/sites/query/queryhtm/npltotal.htm.

Watts, Richard J. 1998. *Hazardous Wastes: Sources, Pathways, Receptors.* New York: John Wiley & Sons.

Section II

Site Investigations and Remediations

3

Chemicals of Concern and Their Properties

3.1 Introduction

Environmental contamination is caused by chemicals that are in some way harmful to human health and the environment. Therefore, to understand how to identify, delineate, and remediate this contamination, it is necessary to have a basic understanding of the processes taking place in the environment. This chapter discusses the principal chemical substances of interest in hazardous waste investigations and remediations, and defines some of their physical properties in the environment.

Chemicals often go by multiple names, either because of their complex molecular structure, ambiguous accepted terminology, or because they are generally known by trade names, abbreviations, or nicknames. In this chapter, we emphasize nomenclature established by the U.S. Environmental Protection Agency (USEPA), although it should be noted that chemical names do not necessarily coincide across various jurisdictions within the USEPA itself, not to mention state and other regulatory programs. This chapter makes note when trade names, abbreviations, or nicknames are used to name a chemical.

The *National Institute for Occupational Safety and Health* (NIOSH) publishes a guidebook that lists hundreds of chemicals, their physical properties, and their various names. This guidebook should be used when one comes across an unfamiliar chemical name. The online version of this guidebook can be accessed at www.cdc.gov/niosh/npg/npgsyn-a.html.

3.2 Categorizing the Chemicals

Chemicals fall into two broad categories: organic and inorganic. Organic compounds contain carbon and are generally (but not necessarily) biological in origin. Inorganic analytes (existing sometimes as compounds, sometimes in elemental form) are not biological in origin, and either do not contain carbon or contain carbon in a minor role.

Millions of chemicals have been identified or manufactured. An undetermined number of these are toxic to human health or the environment. As indicated in Chapter 2, several different lists of chemicals were developed for different sets of regulations. But these lists only contain a tiny portion of the chemicals that exist today. Although on one hand this implies that a huge number of contaminants could be overlooked in the course of a contamination investigation, it is necessary for practical reasons. If an investigation is being conducted on a site at which it is known that a chemical not on one of the lists was used, stored, or manufactured, it should be added to the investigation protocol.

This chapter discusses two of the lists that are used in environmental investigations and remediations: the Target Compound List/Target Analyte List (TCL/TAL) from the Superfund program, and the lists of primary and secondary drinking water standards from the Safe Drinking Water Act. The priority pollutant list from the Clean Water Act is used in environmental investigations and remediations as well. However, with the exception of nine organic compounds, it is a subset of the TCL/TAL and is not discussed in this chapter.

Chemicals that can cause cancer in humans are known as *carcinogens*. The USEPA labels chemicals as known carcinogens, probable carcinogens, or possible carcinogens. Chemicals that are not known or suspected to be cancer causing in humans are known as *noncarcinogens*. The implication of carcinogenicity to environmental investigations and remediations is discussed in Chapter 7.

3.3 The Target Compound List/Target Analyte List

The TCL/TAL currently consists of 173 compounds: 149 organic compounds and 24 inorganic compounds. The following discussion defines the categories and subcategories that most of these substances fall into, and describes the basic chemical compositions of these categories and subcategories. Chemicals of special interest are noted in the text.

3.3.1 Inorganic Analytes

Inorganic analytes found in the environment may be naturally occurring or be manufactured by humans. Table 3.1 lists the inorganic compounds on the TAL. With the exception of cyanide, the inorganic analytes listed in Table 3.1 are elements, although they commonly occur in nature as compounds. A standard laboratory analysis tends not to distinguish between elemental and compound forms of these analytes. In an environmental investigation,

TABLE 3.1

Inorganic Compounds on the Target Analyte List

Aluminum	Cobalt	Potassium
Antimony	Copper	Selenium*
Arsenic*	Iron	Silver*
Barium*	Lead*	Sodium
Beryllium	Magnesium	Thallium
Cadmium*	Manganese	Vanadium
Calcium	Mercury*	Zinc
Chromium*	Nickel	Cyanide

Source: U.S. Environmental Protection Agency.
* RCRA 8 metals.

the total concentration of these analytes is usually what is relevant to the investigator.

Because of their toxicity and their special regulatory status, the so-called RCRA 8 metals (indicated by an asterisk in Table 3.1), are of greater concern than the other metals on the list. Not coincidentally, they tend to be among the major drivers of environmental investigations.

3.3.2 Organic Compounds

One principal method of defining organic compounds is by their volatility. *Volatility* is the ability of a compound to change from liquid to vapor phase. When released into the subsurface, a portion of a volatile compound will evaporate into the gaseous phase. Once in the gaseous phase, the compound can be extremely mobile, traveling great distances through the soils and the man-made structures in the area. The measure of a compound's volatility is its *vapor pressure*, which is the pressure exerted by vapor on a liquid at equilibrium. Vapor pressure varies with temperature, so published vapor pressures must indicate the temperature associated with that vapor pressure.

Vapor pressure is governed by *Henry's law*, which states that under equilibrium conditions, the partial pressure of a volatile chemical above a liquid is proportional to its concentration in the liquid. In mathematical terms

$$p = k_H c$$

where p is the partial pressure of the solute in the gas above the solution, c is the concentration of the solute, and k_H is a constant with the dimensions of pressure divided by concentration.

All organic compounds, to some degree, display volatility; there are no nonvolatile organic compounds. Compounds that readily go into the gaseous phase at temperatures typically found in the environment are known as *volatile organic compounds,* or VOCs. All other organic compounds are *semi-volatile organic compounds,* or SVOCs.

3.3.2.1 Volatile Organic Compounds

Because of their ability to go into gaseous phase, VOCs are the most mobile of compounds, frequently escaping from their point of origin and evaporating, only to condense elsewhere, sometimes far away from their point of origin. Their ability to go into gaseous phase also makes them significant indoor air contaminants, as discussed in Chapter 17.

Table 3.2 lists the 52 VOCs on the TCL, organized by their molecular structures. Fifty of the 52 listed VOCs are volatile *hydrocarbons,* which are organic molecules that contain hydrogen as well as carbon atoms; only carbon tetrachloride and carbon disulfide on the list lack hydrogen in their molecular structures. In volatile hydrocarbon molecules, carbon atoms organize themselves either in chains containing varying numbers of carbon atoms, or in rings, with six carbon atoms forming a hexagon.

Molecular variations occur primarily by the addition of carbon atoms or by the substitution of hydrogen with an ion or a *halogen,* which is a group of elements on the periodic table. The halogens include, in order of atomic number, fluorine, chlorine, bromine, iodine, and astatine. Chlorine is the most common halogen found in compounds on the TCL, followed by fluorine and bromine. No compounds on the TCL contain iodine or astatine.

Most of the molecules shown in Table 3.2 have different molecular formulas. Some, however, have identical molecular formulas but different molecular configurations. These molecules are known as *isomers.* Examples of isomers are shown in Figure 3.1. This figure shows the three dichlorobenzene isomers, all of which appear on the TCL. The numbers preceding the chemical name refer to the "slots" in which the chlorine atoms appear. The 1 slot is defined arbitrarily, and the 2, 3, and 4 slots are defined in relation to the 1 slot. Note that the 5 and 6 slots do not exist; by rotating the molecule, the 1,5 configuration would appear identical to a rotated 1,3 configuration, and the 1,6 configuration would look identical to the 1,2 configuration.

Chained hydrocarbon molecules are known as *aliphatics,* and ringed hydrocarbon molecules are known as *aromatics.* In turn, the aliphatics are divided into three groups, depending on the nature of their carbon bonds: alkanes, alkenes, and alkynes. Each of these groups is described next.

Alkanes are volatile hydrocarbons that have only single bonds in their molecular structure. There are numerous categories of alkanes, four of which are represented on the Target Compound List.

Methane, the simplest alkane, is comprised of one carbon atom and four hydrogen atoms (see Figure 3.2). Replacement of one or more of the hydrogen atoms with other atoms or molecules results in a different molecule, which usually will contain "methane" in its name. For instance, replacement of one of the hydrogen atoms in the methane molecule with a chlorine atom results

TABLE 3.2

Volatile Organic Compounds on the Target
Compound List

Aliphatics

Alkanes

Methanes
   Chloromethane
   Bromomethane
   Bromochloromethane
   Dibromochloromethane
   Dichlorodifluoromethane
   Methylene chloride
   Trichlorofluoromethane
   Chloroform
   Bromoform

Ethanes
   1,1,2,2-Tetrachloroethane
   1,1,2-Trichloroethane
   Trifluoroethane
   1,2-Dichloroethane
   1,1,1-Trichloroethane
   1,1-Dichloroethane
   1,2-Dibromoethane
   Chloroethane
   1,1,2-Trichloro-1,2,2-trifluoroethane

Propanes
   1,2-Dichloropropane
   1,2-Dibromo-3-chloropropane

Hexanes
   Cyclohexane
   Methylcyclohexane

Alkenes

Ethenes
   Tetrachloroethene
   Trichloroethene
   cis-1,2-Dichloroethene
   trans-1,2-Dichloroethene
   1,1-Dichloroethene
   Vinyl Chloride

(Continued)

TABLE 3.2 (CONTINUED)

Volatile Organic Compounds on the Target
Compound List

Propenes
    cis-1,3-Dichloropropene
    trans-1,3-Dichloropropene

Aromatics
Benzene
Toluene
Ethylbenzene
m,p-Xylene
o-Xylene
Isopropylbenzene
Styrene
Chlorobenzene
1,2-Dichlorobenzene
1,3-Dichlorobenzene
1,4-Dichlorobenzene
1,2,3-Trichlorobenzene
1,2,4-Trichlorobenzene

Other VOCs
Methyl tertiary-butyl ether
Methyl acetate
1,4-Dioxane
Carbon tetrachloride
Carbon disulfide

Ketones
Acetone
2-Butanone
4-Methyl-2-pentanone
2-Hexanone

Source: U.S. Environmental Protection Agency.

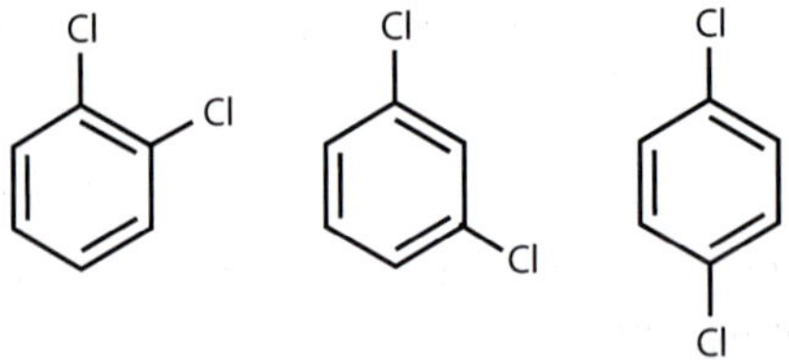

FIGURE 3.1
The three dichlorobenzene isomers. From left to right: 1,2-dichlorobenzene, 1,3-dichloroben-
zene, and 1,4-dichlorobenzene.

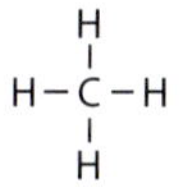

```
        H
        |
    H − C − H
        |
        H
```

FIGURE 3.2
A methane molecule.

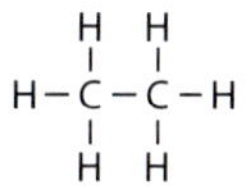

```
      H   H
      |   |
  H − C − C − H
      |   |
      H   H
```

FIGURE 3.3
An ethane molecule.

in the formation of chloromethane, a VOC on the TCL. Replacing all four hydrogen atoms with two chlorine atoms and fluorine atoms results in the formation of dichlorodifluoromethane, which is also on the TCL.*

Ethanes are alkanes that contain two carbon atoms linked by a single bond. In ethane, the six available slots are filled six hydrogen atoms (see Figure 3.3). As with methane, replacement of one or more hydrogen atoms with a halogen yields several compounds that are on the TCL. For instance, 1,2-dichloroethane has two chlorine atoms in place of two of the hydrogen atoms (the numbers in front of the chemical name refer to the locations of the chlorine atoms in the molecule.). 1,1,1-Trichloroethane, commonly known as TCA, has three chlorine atoms in place of three hydrogen atoms, and so on.

Propanes are alkanes that contain three carbon atoms linked by a single bond, and *hexanes* are alkanes that contain six carbon atoms linked by a single bond (see Figure 3.4). In general, the greater the molecular weight, the less volatile the molecule. Saturated molecules (molecules containing single bonds) with more carbon atoms, or with heavier atoms substituting for hydrogen atoms, are less likely to be sufficiently volatile to be classified as VOCs, and more likely to be classified as SVOCs (see Section 3.3.2.2).

Alkenes are volatile hydrocarbons that have a double bond linking their carbon atoms. The ethenes represented on the Target Compound List fall into one of two categories. *Ethylene* (also known as *ethene*), the simplest alkene, is comprised of two carbon atoms and four hydrogen atoms (see Figure 3.5). The six chlorinated ethylenes on the Target Compound List are worthy of special note. They are related compounds, differing from each other only by the number and placement of the chlorine atoms around the carbon atoms. The molecule with four chlorine atoms, is known alternatively as perchloroethene, perchloroethylene, tetrachloroethene, or tetrachloroethylene,

---

* Dichlorodifluoromethane, because of its extremely low vapor pressure, is an excellent refrigerant. Unfortunately, as its name implies, it is a chlorofluorocarbon, or CFC, known for its imperilment of the earth's ozone layer. This compound (commonly known as Freon 12) as well as its sister CFCs were placed under a worldwide production ban in the 1990s.

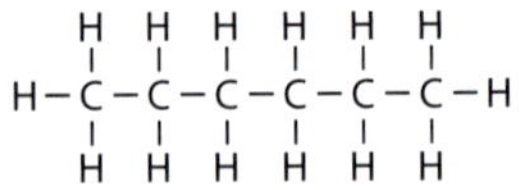

```
      H   H   H   H   H   H
      |   |   |   |   |   |
  H - C - C - C - C - C - C - H
      |   |   |   |   |   |
      H   H   H   H   H   H
```

FIGURE 3.4
A hexane molecule.

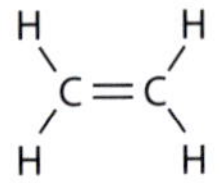

```
   H          H
    \        /
     C  =  C
    /        \
   H          H
```

FIGURE 3.5
An ethylene molecule.

PCE, or, by the dry-cleaning industry, as perc. Most dry-cleaning establishments, use perc in the dry-cleaning process because of its exceptional ability to remove dirt from garments (see Chapter 5). Its sister compound, known as trichloroethene, trichloroethylene, "trichlor," or TCE, was also a highly effective cleaner before its use declined due to environmental concerns. The other compounds in the series, the three dichloroethenes (containing two chlorine atoms and two hydrogen atoms), and vinyl chloride (one chlorine and three hydrogen atoms), derive from the *dechlorination* (substitution of a hydrogen atom for a chlorine atom) of PCE or TCE. *Propenes* contain three carbon atoms, with one double bond and one single bond (see Figure 3.6). This leaves six slots for halogens or a molecule with the appropriate valence. The two propenes on the Target Compound List each contain two chlorine atoms in their molecular structure.

Alkynes are compounds that possess a carbon-to-carbon triple bond. An example of an alkyne is shown in Figure 3.7. There are no alkynes on the Target Compound List.

As mentioned earlier, *aromatics* are molecules comprised of six carbon atoms that form a ring. *Benzene* is the simplest aromatic compound. It has hydrogen atoms filling all of its available slots around the ring. Benzene, along with toluene, ethylbenzene, and the two xylene isomers, form a group of chemicals known as BTEX (see Figure 3.8). Most of the other benzene-related compounds on the Target Compound List contain one to three chlorine atoms around the benzene ring.

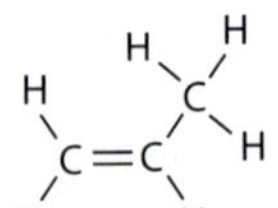

```
                   H    H
                    \  /
              H      C
               \    / \
                C = C   H
               /    \
              H      H
```

FIGURE 3.6
A propene molecule.

$$R-C\equiv C-R'$$

FIGURE 3.7
An alkyne molecule, where R represents any molecule or atom with the appropriate valence.

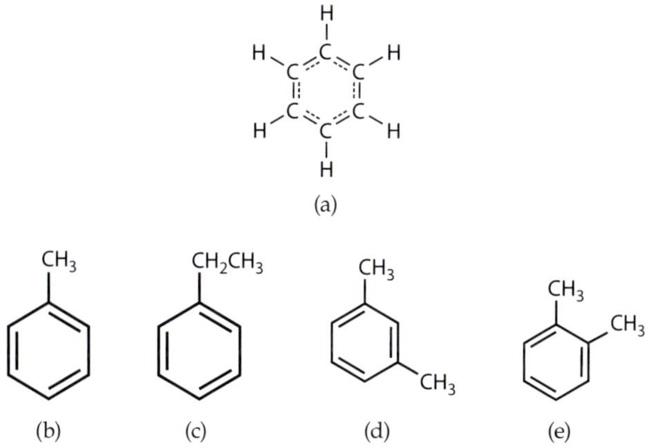

(a)

(b) (c) (d) (e)

FIGURE 3.8
(a) Benzene molecule. (b) Toluene molecule, with implied carbon atoms in the benzene ring. (c) Ethylbenzene molecule. (d and e) m-xylene and o-xylene, two of the three xylene isomers.

Benzene itself is a known carcinogen, and, along with PCE and TCE, is considered to be one of the most potentially harmful VOCs on the Target Compound List. These three compounds drive an inordinate amount of subsurface investigations.

Ketones contain a carbonyl group, which is an oxygen atom double bonded to a carbon atom, which is in turn bonded to two other carbon-based molecules. Acetone, which is shown in Figure 3.9, is the simplest ketone. It is a common industrial cleaner and is often used to clean laboratory equipment.

The final group discussed under VOCs only contains one compound on the Target Compound List. *Ethers* (see Figure 3.10) contain an ether group, which consists of an oxygen atom attached to a combination of two of the following: one or two alkyl groups (which are alkanes minus one hydrogen atom) and one or two aryl groups (which are benzene rings minus one hydrogen atom). Methyl tert-butyl ether (MTBE), the only ether on the VOC portion of the Target Compound List, is an additive to gasoline that is discussed in Section 3.4.1.2.

3.3.2.2 Semivolatile Organic Compounds

Semivolatile organic compounds, or SVOCs, are generally larger molecules than VOCs and are less likely to go into the gaseous phase, hence their name. Table 3.3 lists the SVOCs on the Target Compound List, organized by their molecular structure. It should be noted that some SVOCs fall into the same

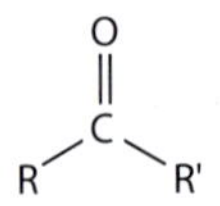

$$
\underset{R}{\overset{\displaystyle \overset{O}{\underset{\|}{C}}}{\diagdown}} \underset{R'}{}
$$

FIGURE 3.9
A ketone molecule.

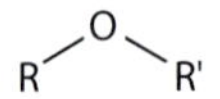

$$
R \diagup \overset{O}{\diagdown} R'
$$

FIGURE 3.10
An ether molecule.

categories as VOCs, such as methanes, ethanes, benzenes, ketones, and ethers. The major categories not shared with VOCs are discussed later.

Phthalates (see Figure 3.11) are esters commonly used as plasticizers, which are substances used in the manufacture of plastics. Of particular note among the phthalates listed in Table 3.3 is bis(2-ethylhexyl) phthalate, also known as DEHP. It is a common laboratory contaminant, since PVC (polyvinyl chloride), a common plastic, can emit DEHP as an off-gas, especially when the equipment is new.

The largest grouping of SVOCs on the Target Compound List is *polycyclic aromatic hydrocarbons*, alternately abbreviated as PAHs (polyaromatic hydrocarbons) or PNAs (polynuclear aromatics). As their name implies, the compounds are in the aromatics family, comprised of at least two interconnected benzene rings. Naphthalene, shown in Figure 3.12, is the simplest of the PAHs. It consists of two benzene rings and no other atoms. Naphthalene and its sister compound 2-methylnaphthalene are often produced from the distillation of coal tar and the refining of petroleum, and are often found in a variety of petroleum products.

The other PAH compounds have various atoms and molecules attached to the dual benzene rings. They are derived either from the distillation of coal tar or from incomplete combustion of organic material, especially fossil fuels. Of particular concern are the last nine PAHs listed in Table 3.3, beginning with benzo(a)anthracene. They typically are comprised of a benzene ring attached to one of the PAHs with similar names (anthracene, pyrene, etc.). These compounds taken together are known as the *carcinogenic PAHs*. As their name implies, they are known carcinogens, which elevate them to a level of concern far above the other SVOCs on the Target Compound List. Benzo(a)pyrene, shown in Figure 3.13, is considered the most toxic of the carcinogenic PAHs (sometimes abbreviated CaPAHs).

Phenyls are comprised of a phenyl ring, which is a benzene ring with a molecule replacing one of the ring's hydrogen atoms (see Figure 3.14). 1,1'-Biphenyl and N-nitrosodiphenylamine contain two attached phenyl rings. When multiple phenyl rings have multiple chlorine atoms attached

TABLE 3.3

Semivolatile Organic Compounds on the Target
Compound List

Methanes/Ethanes
Hexachloroethane
Bis(2-chloroethoxy) methane
2,2′-Oxybis(1-choloropropane)

Benzenes
Hexachlorobenzene
2,6-Dinitrotoluene
2,4-Dinitrotoluene
Nitrobenzene
1,2,4,5-Tetrachlorobenzene
Benzaldehyde

Ketones
Isophorone
Acetophenone

Ethers
Bis(2-chloroethyl) ether
4-Chlorophenyl-phenyl ether
4-Bromophenyl-phenylether

Phthalates
Bis(2-ethylhexyl) phthalate
Dimethylphthalate
Diethylphthalate
Di-n-butylphthalate
Butylbenzylphthalate
Di-n-octylphthalate

Polycyclic Aromatics
Naphthalene
2-Methylnaphthalene
2-Chloronaphthalene
Acenaphthene
Acenaphthylene
Anthracene
Fluorene
Fluoranthene
Pyrene
Phenanthrene
Benzo(a)anthracene
Benzo(b) fluoranthene

(Continued)

TABLE 3.3 (CONTINUED)

Semivolatile Organic Compounds on the Target
Compound List

Benzo(k) fluoranthene
Benzo(a) pyrene
Benzo(g,h,i) perylene
Chrysene
Dibenzo(a,h) anthracene
Indeno(1,2,3,-cd) pyrene

Phenyl
2-Nitroaniline
3-Nitroaniline
4-Nitroaniline
4-Chloroaniline
1,1′-Biphenyl
N-Nitroso-di-n phenylamine

Phenols
Phenol
4-Chloro-3-methylphenol
2,4-Dichlorophenol
2-Nitrophenol
2,4-Dimethylphenol
2-Chlorophenol
2-Methylphenol
2,4-Dinitrophenol
4-Nitrophenol
2,3,4,6-Tetrachlorophenol
Pentachlorophenol
2,4,6-Trichlorophenol
2,4,5-Trichlorophenol
4-Methylphenol
4,6-Dinitro-2-methylphenol

Others
Atrazine
Caprolactam
N-Nitrosodi-n-propylamine
Carbazole
Dibenzofuran
3,3′-dicholorobenzidine
Hexachlorocyclopentadiene
Hexachlorobutadiene

Source: U.S. Environmental Protection Agency.

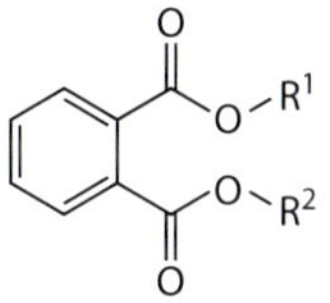

FIGURE 3.11
A phthalate molecule.

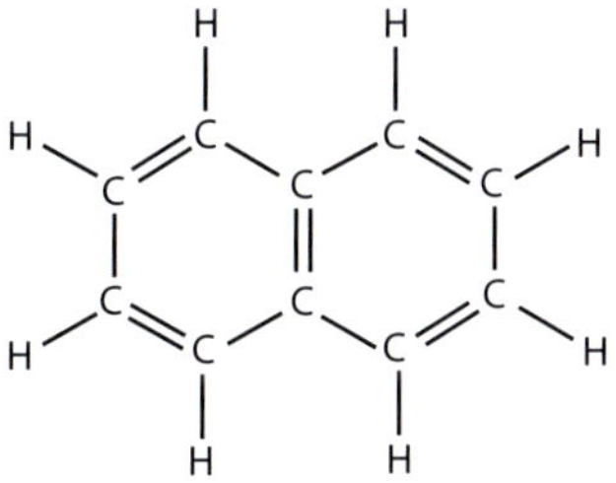

FIGURE 3.12
A naphthalene molecule.

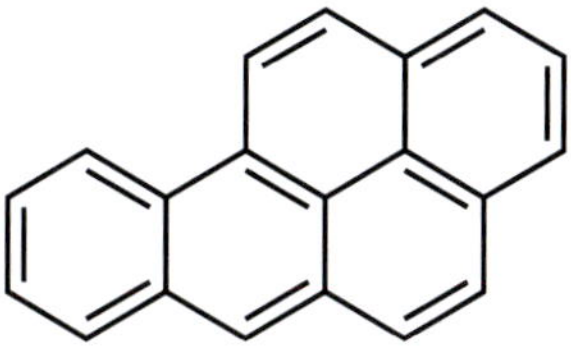

FIGURE 3.13
A benzo(a)pyrene molecule, comprised of five interconnected benzene rings.

to them, they fall into a special category of SVOCs known as *polychlorinated biphenyls*, or *PCBs*, which are discussed below.

Phenols consist of a hydroxyl molecule (OH⁻) attached to a benzene ring (see Figure 3.15). This group of compounds is sometimes referred to as *acid extractables*, or *AEs*, due to the method by which samples are prepared in the laboratory for analysis. In contrast, the other SVOCs on the Target Compound List are known as *base-neutral compounds*, or *BNs*, again due to the method by which a laboratory will prepare the samples for analysis.

3.3.2.3 Pesticides and Polychlorinated Biphenyls (PCBs)

The other two primary categories of organic compounds on the TCL are defined by their function rather than their chemistry. These two categories are *pesticides* and *polychlorinated biphenyls*, more commonly known as PCBs.

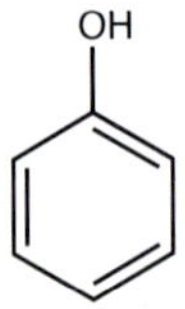

FIGURE 3.14
A phenyl molecule.

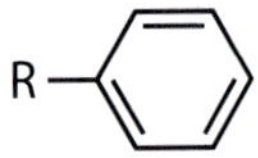

FIGURE 3.15
A phenol molecule.

The TCL pesticide list (see Table 3.4) contains organic compounds with a specific function: to kill undesirable fauna, mainly insects and rodents. These pesticides, while having different chemical compositions, have one thing in common: they all have chlorine atoms in their molecular structures. Hence their family name: *organochlorine pesticides*.

TABLE 3.4

Organic Pesticides on the Target
Compound List

alpha-BHC
beta-BHC
gamma-BHC (Lindane)
delta-BHC
Heptachlor
Aldrin
Heptachlor epoxide
Endosulfan I
Dieldrin
4,4′-DDE
Endrin
Endosulfan II
4,4′-DDD
Endosulfan sulfate
4,4′-DDT
Methoxychlor
Endrin ketone
Endrin aldehyde
alpha-Chlordane
gamma-Chlordane
Toxaphene

Source: U.S. Environmental Protection Agency.

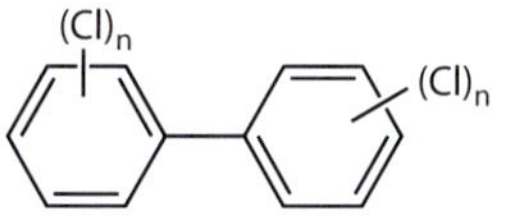

FIGURE 3.16
A PCB molecule. The number of chlorine atoms in its molecular structure varies based on congener.

The best known pesticide, dichlorodiphenyltrichloroethane, is commonly known by its initials: DDT. As its name implies, it is comprised of two phenyl molecules attached to a trichloroethane molecule (that is, an ethane molecule with three chlorine atoms). DDT and many of the other pesticides on the Target Compound List are banned in the United States. However, because of their persistence in the environment, they still are often encountered in the locations where they were applied.

Prior to the rise in the usage of organochlorine pesticides during and after World War II, the pesticide of choice was lead arsenate, an inorganic pesticide (chemical formula $PbHAsO_4$). When investigating for the presence of pesticides on current or former farmland, lead and arsenic are often added to the laboratory analyses, especially if the land was used for farming before World War II.

PCBs, which were outlawed in 1979 under the Toxic Substances Control Act (TSCA), had widespread usage in electrical equipment due to their stability under high temperature and pressure conditions. They were also commonly used in hydraulic oils and other oils for which stability was a priority.

PCBs are composed of two connected phenyl rings, which are similar to benzene rings. Attached to the two phenyl rings (biphenyl) are a varying number of chlorine atoms (see Figure 3.16).

There are 209 types of PCBs, known as *congeners*. However, PCBs are commonly grouped as Aroclors, which was a trade name under which Monsanto Corporation manufactured PCBs, but which, over time, has become the generic name for the various PCBs. The last two digits in the name indicate the chlorine content of the molecule. For instance, Aroclor-1254 is 54% chlorine atoms by weight. The nine Aroclors on the Target Compound List are shown in Table 3.5.

3.4 Other Chemicals of Interest

3.4.1 Petroleum and Petroleum-Related Compounds

Petroleum does not have a set chemical formula. Rather, it is a complex mixture of chemicals. It occurs naturally in the form of crude oil and generally is

TABLE 3.5

Targeted PCBs (Aroclors)

Aroclor-1016
Aroclor-1221
Aroclor-1232
Aroclor-1242
Aroclor-1248
Aroclor-1254
Aroclor-1260
Aroclor-1262
Aroclor-1268

Source: U.S. Environmentl Protection Agency.

not toxic to humans. It is, however, dangerous to flora and fauna, especially birds and fish, when spilled. Certain petroleum additives can make it toxic to humans, as described next.

3.4.1.1 Chemical Composition of Petroleum

Two things that all types of petroleum have in common are carbon atoms and hydrogen atoms, hence the name hydrocarbons. The simplest type of petroleum is methane, which contains one carbon atom, as described earlier. Most types of petroleum, however, involve multiple carbon atoms, arranged in chains that consist of two or more carbon atoms bonded together. The hexane molecule shown in Figure 3.3 is a straight-chain hydrocarbon containing six carbon atoms. Adding more carbon atoms to the chain would create ever-larger hydrocarbon molecules (see Figure 3.17). The notation commonly used to describe hydrocarbon chains is C_n, where n is the number of carbon atoms in the chain.

Crude oil goes through a refining process to become a substance that is useful to humans. In the refining process, impurities and other undesirable chemicals are removed from the crude, and desirable chemicals are added to the crude, to form a petroleum product designed with specific uses in mind. Some of the more common types of refined petroleum products are described next.

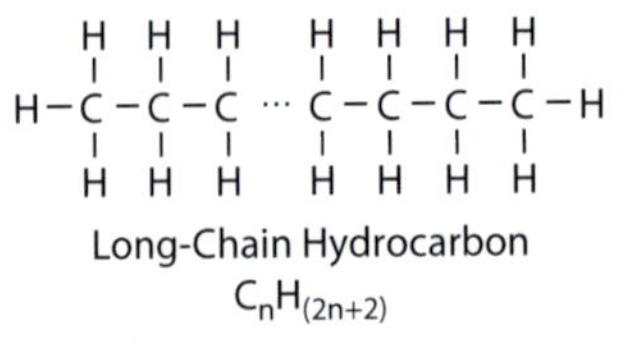

Long-Chain Hydrocarbon

$$C_nH_{(2n+2)}$$

FIGURE 3.17
General formula for a straight-chain hydrocarbon.

3.4.1.2 Gasoline

The most common use for petroleum is in the manufacture of fuels designed for the rapid delivery of power. These usually contain flammable components whose bonds will break apart and yield energy instantly when injected into the internal combustion engine of a vehicle or a similar device. These volatile components also render the fuel flammable, thus hazardous to work with and store as well as toxic to human beings.

The most common flammable components added to gasoline are the BTEX compounds, discussed earlier in the chapter. The BTEX compounds, when ignited, give the internal combustion engine the punch that is so familiar to those of us who have pressed down on the accelerator. Until the mid-1970s, lead, typically in the form of tetra-ethyl lead, was added to the gasoline to provide that power. The practice of adding lead to gasoline was discontinued under the Clean Air Act to reduce the amount of lead in the air.

Starting in the 1980s and accelerating into the 1990s under the Clean Air Act Amendments of 1990 was the use of oxygenates in gasoline. The role of oxygenates is to enable the gasoline to burn more cleanly and create less carbon monoxide emissions, thereby reducing smog. The original oxygenate of choice was MTBE. Although it does its job well, MTBE use in the United States has been recently declining due to environmental and health concerns. It is slowly being replaced by ethanol, which is a nonpetroleum, simple alcohol derived from corn.

3.4.1.3 Nonvolatile Fuels

Nonvolatile fuel oils are designed for delivery of power as well as stability in the environment. They tend to be numbered by their degree of volatility.

No. 1 fuel oil, more commonly known as *kerosene*, is a less refined petroleum product than gasoline. It is the basis for jet fuel; however, most kerosene products are not volatile, and therefore are more stable than gasoline or jet fuel, and safer to use and store.

Diesel fuel, or no. 2 fuel oil, is widely used in internal combustion engines of larger motor vehicles, such as large cars, trucks, buses, and trains. These vehicles do not need the acceleration desired in a standard automobile, although turbocharging automotive diesel engines is becoming a popular way of enhancing the engine's power. This fuel is very stable in the environment, which is why it is widely used to heat buildings.

Of the commonly used fuel oils, the heaviest and most stable is known as no. 6 fuel oil. It is often called bunker oil, although that phrase is generally meant to describe the oil used on ships. It is so viscous (thick) that it usually has to be heated to flow, which has important implications for spills of no. 6 fuel oil.

No. 4 fuel oil and no. 5 fuel oil are mixtures of no. 2 fuel oil and no. 6 fuel oil. No. 3 fuel oil technically exists, but is rare.

3.4.1.4 Engineered Oils

In addition to the fuel oils, there are literally thousands of petroleum products that are engineered with other uses in mind. Many petroleum products are designed for stability under extreme conditions of heat or pressure, or as lubricants. Some of the more common forms of engineered oils are

Mineral spirits, or mineral oil

Lubricating oil

Hydraulic oil

Cutting oil

Crankcase

Grease

The composition of the literally thousands of petroleum products cannot be generalized. Each product contains a unique mix of various types of petroleum and petroleum additives. The only ways to understand the chemical composition of a petroleum product, even a fuel oil, is to either analyze the product in a laboratory or read its material safety data (MSD) sheet (see Chapter 2).

3.4.2 Dioxin

Dioxin refers to a family of chemical compounds that are generally considered to be among the most hazardous to human health and the environment. Generally comprised of two benzene rings connected by oxygen "bridges," they are unwanted by-products in the manufacture of organic pesticides and by incomplete combustion. 2,3,7,8-Tetrachlorodibenzo-*p*-dioxin (TCDD) is the most potent compound of the series and became infamous nationwide as a contaminant in Agent Orange, an herbicide used in the Vietnam War.

3.4.3 Drinking Water Contaminants

The Safe Drinking Water Act established primary maximum contaminant levels (MCL) and maximum contaminant level goals (MCLG) for various biological, chemical, and radiological contaminants, and secondary National Secondary Drinking Water Regulations (NSDWRs or secondary standards) for 15 other chemicals or chemical properties. These chemicals are important to environmental investigations and remediations when drinking water or an aquifer with the potential of being used as a source of drinking water is affected.

Table 3.6 lists the organic chemicals for which MCLs and MCLGs have been established under the Safe Drinking Water Act. The reader will note

TABLE 3.6

Organic Chemicals Regulated under the Safe
Drinking Water Act

VOCs
Dichloromethane
1,2-Dichloroethane
1,1,1-Trichloroethane
1,1,2-Trichloroethane
1,2-Dichloropropane
Tetrachloroethylene
Trichloroethylene
1,1-Dichloroethylene
Cis-1,2-Dichloroethylene
Trans-1,2-Dichloroethylene
Vinyl chloride
Benzene
Ethylbenzene
Toluene
Xylenes (total)
Chlorobenzene
o-Dichlorobenzene
p-Dichlorobenzene
1,2,4-Trichlorobenzene
Styrene
Carbon tetrachloride

SVOCs
Atrazine
Benzo(a)pyrene (PAHs)
Di(2-ethylhexyl)phthalate
Hexachlorobenzene
Hexachlorocyclopentadiene
Di(2-ethylhexyl)adipate
Ethylene dibromide

Herbicides
Alachlor
Dinoseb
Diquat
Endothall
Glyphosate
Picloram
Simazine

(Continued)

TABLE 3.6 (CONTINUED)

Organic Chemicals Regulated under the Safe
Drinking Water Act

Pesticides/PCBs
Carbofuran
Chlordane
2,4-D
Dalapon
1,2-Dibromo-3-chloropropane (DBCP)
Endrin
Heptachlor
Heptachlor epoxide
Lindane
Methoxychlor
Oxamyl (Vydate)
Pentachlorophenol
Toxaphene
2,4,5-TP (Silvex)
Polychlorinated Biphenyls (PCBs)

Others
Dioxin (2,3,7,8-TCDD)
Acrylamide
Epichlorohydrin

Source: U.S. Environmental Protection Agency.

that most but not all of the chemicals listed in the table are on the Target
Compound List.

Table 3.7 lists the inorganic chemicals, biological agents, radionuclides,
and chemical properties for which MCLs and MCLGs have been established
under the Safe Drinking Water Act. The metals and cyanide listed here also
appear on the Target Analyte List. The listings for the other categories do
not appear on the Target Analyte List, and, with the exception of turbidity,
nitrates, and nitrites, are generally not considered when conducting environmental investigations and remediations.

Table 3.8 lists the NSDWRs, which are chemicals or chemical properties that
may affect the taste, odor, or color of the drinking water, or cause cosmetic
effects, such as skin or tooth discoloration. The USEPA recommends but does
not require regulation of these parameters, except for publically owned treatment works (see Chapter 2). Some familiar names appear on this list.

TABLE 3.7

Inorganic Chemicals and Other Parameters
Regulated under the Safe Drinking Water Act

Microorganisms
Cryptosporidium
Total coliforms (including fecal coliform and *E. coli*)
Giardia lamblia
Heterotrophic plate count
Legionella
Turbidity
Viruses (eteric)

Metals
Antimony
Arsenic
Barium
Beryllium
Cadmium
Chromium (total)
Copper
Lead
Mercury (inorganic)
Selenium
Thallium

Disinfection Byproducts
Bromate
Chloramines (as Cl_2)
Chlorine (as Cl_2)
Chlorite
Haloacetic acids (HAAS)
Total trihalomethanes

Radionuclides
Alpha particles
Beta particles and photon emitters
Radium 226 and Radium 228 (combined)

Others
Cyanide (as free cyanide)
Asbestos (fiber >10 micrometers)
Fluoride
Nitrate (measured as nitrogen)
Nitrite (measured as nitrogen)

TABLE 3.8

National Secondary Drinking Water
Regulations Parameters

Metals	Others
Aluminum	Chloride
Copper	Color
Iron	Corrosivity
Manganese	Fluoride
Silver	Foaming agents
Zinc	Odor
	pH
	Sulfate
	Total dissolved solids (TDS)

Source: U.S. Environmental Protection Agency.

3.4.4 Synthetic Organic Contaminants

The other millions of chemicals known to exist do not need to be ignored during an environmental investigation or remediation, since the analyses performed to detect the organic chemicals on the various lists also pick up other chemicals with similar physical properties, especially boiling point. A laboratory can attempt to identify nontargeted compounds based on the physical properties that they exhibit during the analysis. Such compounds are known as *tentatively identified compounds*, or TICs. No standards exist for individual TICs; however, many jurisdictions have standards for total organic compounds in the aggregate. The term used, which includes "targeted" compounds as well as TICs, is total *synthetic organic contaminants* (SOCs).

References

Budavari, Susan, O'Neil, Maryadele J., Smith, Ann, Heckelman, Patricia E., and Kinneary, Joanne F. 1996. *The Merck Index*, 12th ed. Whitehouse Station, NJ: Merck & Co.

International Union of Pure and Applied Chemistry. 1993. *Quantities, Units, and Symbols in Physical Chemistry*, 2nd ed. New York: Blackwell Science.

Lagrega, Michael D., Buckingham, P. L., and Evans, J. E. 2001. *Hazardous Waste Management*, 2nd ed. New York: McGraw-Hill.

Meyer, Eugene. 1989. *Chemistry of Hazardous Materials*, 2nd ed. New York: Prentice Hall.

Schecter, Arnold, Birnbaum, Linda, Ryan, John J., and Constable, John D. 2006. Dioxins: An overview. *Environmental Research,* 101(3): 419–428.

Watts, Richard J. 1997. *Hazardous Wastes: Sources, Pathways, Receptors.* New York: John Wiley & Sons.

4

Fate and Transport in the Subsurface

In a laboratory, chemicals behave in predictable ways due to the controlled nature of a laboratory. Although the basic principles still apply, in the uncontrolled environment, chemical behavior is far more complicated. Physical, chemical, and biological forces influence the occurrence and movement of a chemical in the environment. This chapter describes the role of the physical framework of the environment—its geology—in determining the fate and transport of chemicals.

4.1 Surface Transport of Chemicals

At the surface, chemicals are transported either through the air or by surface water. When transported in the air, the chemicals may be in solid or gaseous phase. When in solid phase, a chemical is typically attached to a particulate, be it a speck of dust or a grain of soil. Airborne, the particulate will transport the chemical until it settles on the ground. Particles can travel great distances before settling on the ground, as is evidenced by large volcanic eruptions, which have resulted in the dispersion of fine particles being transported around the world several times over.

When transported in surface water, such as a river, the particle will move in the direction of river flow, that is, downstream toward its endpoint, usually a lake, sea, or other larger water body. Suspended particles can travel significant distances in moving water. In the Hudson River PCB Superfund case, suspended particles containing PCBs traveled over 100 miles from their point of origin. Such cases, however, are unusual, and long-distance migration via surface water is more the exception than the rule when considering the impact of other properties on the site.

A river or stream will transport chemicals in one of three ways. Soluble constituents will travel in as solutes dissolved in water at the same rate as the water. Nonaqueous phase liquids, such as many petroleum products, will travel downstream but not necessarily at the same rate as the water. Constituents sorbed to silt, sand, or organic particles suspended in the water will also travel downstream but generally at a slower rate than the flowing water. Dissolved chemicals and chemicals sorbed to particles will interact with other chemicals in the water as well as with chemicals sorbed to other suspended particles or the underlying sediments.

Chemicals in the gaseous phase can travel even greater distances in air than in water. Even without the energy of, say, a volcano, chemicals can travel hundreds of miles through the air. Chemicals released in industrial settings other than power plants tend to travel more modest distances in the air, usually not much further than the general vicinity of the plant. These chemicals may eventually deposit themselves nearby, often by condensing into liquid phase.

4.2 Geology of the Subsurface

Before explaining chemical fate and transport in the subsurface, a number of geologic and physical principles must be identified and described. First is a brief description of the geology that chemicals typically encounter in the subsurface.

4.2.1 Bedrock and Soils

Subsurface materials generally fall into two categories: bedrock and soils. Bedrock consists of either igneous, sedimentary, or metamorphic rock. *Igneous* rocks originate from cooling magma that has welled up from deep within the earth's crust or from cooling lava from volcanoes at the surface of the earth's crust. *Sedimentary* rocks are formed by the consolidation of sediments by the removal of water from their pore structure, usually due to the pressure of overlying sediments or dessication at or near the surface, or through chemical or biological activity, as in the case of limestone. *Metamorphic* rocks are created by changing the molecular structure of its component minerals by the application of heat and pressure to igneous or sedimentary rocks.

Soils are unconsolidated sedimentary deposits, and they are classified primarily by grain size. There are several different systems of soil classification, but the basic categories are the same (see Table 4.1 for one method of soils classification). *Clay*-sized particles are too small to be observed with the naked eye. *Silt* is comprised of very small grains that can be sensed by touch and observed without use of a microscope, despite their small size. *Sand* is comprised of grains that are easily observed with the naked eye. Sands usually carry secondary descriptors based on grain size: very fine, fine, medium, coarse, and very coarse. The next grain size up from sands is *gravel*. Larger particles are known as *boulders*.

The description of soils with more than one soil type can often be subjective. For instance, soil with both sand-sized and silt-sized particles can be classified as "silty sand" or "sandy silt." The primary classification should reflect the dominant soil type, which can be difficult to discern without quantitative

TABLE 4.1

Wentworth's Clastic Scale

Major Clasts	Specific Clasts	Sizes
	Boulder	>256 mm
Gravel	Cobble	64–256 mm
	Pebble	4–64 mm
	Granule	2–4 mm
	Sand	1/16–2 mm
Mud	Silt	1/256–1/16 mm
	Clay	<1/256 mm

Source: Anderson, John R., 2008, The World of Geology, http://facstaff.gpc.edu/~janderso.

data. Therefore, the most reliable way to determine the proper classification for soil with more than one soil type is by collecting quantitative data, which is typically done by performing a *grain size analysis*. The simplest and most common grain size analysis is known as a sieve test (see Figure 4.1). In a sieve test, the soil is sent through a series of filters of progressively smaller aperture. The soils retained at each level of filters are then weighed. The level with the largest quantity of soil becomes the primary soil descriptor.

Other soil descriptors rely on visual classifications, the most obvious being color. Sands and gravels are also described by the angularity of their grains. These secondary classifications identify the sand or gravel grains as angular, subangular, or rounded (see Figure 4.2). Soils are also described by the uniformity of their grains. Soils with uniform grain size are labeled *well sorted*, and soils with a hodgepodge of soil grain sizes are labeled *poorly sorted*.

The other important visual soil descriptor is its moisture content. Soils that are *saturated* contain only liquids and no gases in their pore spaces. *Unsaturated* soils must contain some gas in their pore spaces, but also may contain liquids. Unsaturated soils that contain some liquids usually are described as "moist" or wet to the touch.

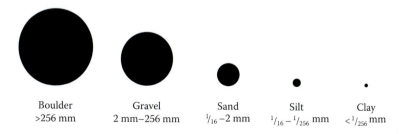

Boulder	Gravel	Sand	Silt	Clay
>256 mm	2 mm–256 mm	$\frac{1}{16}$–2 mm	$\frac{1}{16}$ – $\frac{1}{256}$ mm	< $\frac{1}{256}$ mm

FIGURE 4.1

Comparative grain sizes of major soil types.

4.2.2 Porosity and Permeability

Two basic soil properties that cannot be determined by qualitative observation but are nonetheless critical to the understanding of chemical transport in the subsurface are porosity and permeability. *Porosity* is the capacity of a soil to hold liquid or air. It is a dimensionless ratio, measured by dividing the volume of pore spaces by the total volume. As shown in Figure 4.2, sand with uniformly sized grains will have the maximum porosity. Consultants often use a 30% porosity in equations for porous soil for which no quantitative data are available.

Since they are comprised of interlocking mineral crystals rather than discrete particles, igneous and metamorphic rocks do not have pore spaces (primary porosity). However, differential stresses in the subsurface can create fractures in these rocks, creating what is referred to as *secondary porosity.* Intersecting fractures in bedrock can contain and transport significant quantities of water and solutes (see Figure 4.3).

Limestone, a type of sedimentary rock, can dissolve over time, creating large dissolution cavities that can form caverns. Dissolved limestone formations form *karst topography,* which is characterized by uneven land surfaces.

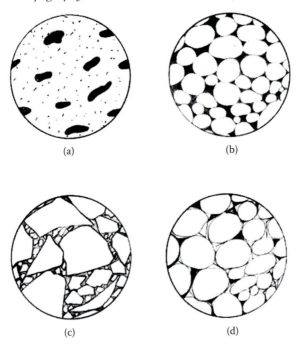

(a)   (b)

(c)   (d)

FIGURE 4.2
Variations in grain size and angularity. (a) Poorly sorted sand grains. (b) Poorly sorted well-rounded sand grains. (c) Poorly sorted, angular sand grains. (d) Well-sorted, rounded sand grains. (From Anderson, John R., 2008, The World of Geology, http://facstaff.gpc. edu/~janderso/images/physical/grdwF2.jpg.)

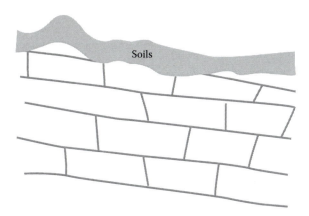

FIGURE 4.3
Complex bedrock fracture pattern.

Underground streams can flow through the caverns and may eventually undermine the overlying surface, causing it to collapse. Such collapsed structures, called *sinkholes,* are a major hazard in areas with karst topography.

Permeability describes the interconnectivity of pore spaces in soil. Soils with good permeability allow for the efficient transport of liquids in a manner similar to good roadways allowing for the efficient transport of motor vehicles. Soil with good permeability will have interconnected "roadways" and lots of them. Soil with poor permeability will have few roadways or many roadways with dead ends. Just as porosity defines the quantity of water that can be found in a soil or rock formation, permeability dictates how easy it is for the water and accompanying dissolved chemicals to move around.

4.3 Chemical Fate and Transport in the Subsurface

4.3.1 Physical State of Chemicals

The ability of a compound to exist in two phases, such as liquid and gas, at the same time is known as its *partition coefficient.* There are many types of partition coefficients. A compound's *vapor–liquid partition coefficient* is the ratio of the concentration of a compound in air to the concentration of the compound in liquid at equilibrium. Other partition coefficients are discussed later in this chapter.

A chemical in the gaseous phase that is released at the surface is unlikely to impact the subsurface unless subsequent events cause the chemical to condense and dissolve into the liquid phase, or it is released directly into a surface water body. A chemical released in a solid phase will not impact the subsurface immediately, unless it becomes dissolved in water before being cleaned up. Even for chemicals in the gaseous or solid phases, water is usually involved in their transport into the environment. This holds even more for chemicals released into the environment in the liquid phase.

Solubility is the ability of a compound, known as the solute, to dissolve in another compound, which is known as the solvent. It is usually expressed as a simple ratio of the molecules that will dissolve into solution divided by the molecules of solvent. In almost all cases, and unless otherwise stated, the solvent in environmental investigations is water.

Chemical literature often categorizes compounds as "soluble" or "insoluble." However, these qualitative labels can be misleading. Maximum allowable concentrations of compounds on the various compound lists typically are expressed in parts per million, parts per billion, or, in a few cases, even parts per trillion. Standard scientific language is usually geared toward manufacturing processes, where chemical concentrations are often expressed as a percentage of the total weight of the product. Because of this dichotomy, chemicals that are considered to be insoluble in water by standard terminology are usually quite soluble at the concentrations under consideration in an environmental investigation. For that reason, the term insoluble generally is not used in an environmental investigation, except as a qualitative description of a given compound.

Solubility can also be expressed in terms of a chemical's *octanol–water partition coefficient*, which describes whether a contaminant is hydrophilic (attracted to water) or hydrophobic (repelled in water). It is expressed by the ratio

$$K_{ow} = C_o/C_w$$

where C_o is the concentration of the chemical in octanol (a petroleum with eight carbon atoms) and C_w is the concentration of the chemical in aqueous phase (dissolved in water.)

A chemical in the liquid phase will either be dissolved in a solvent or will be a pure-phase chemical existing in liquid phase at ambient temperatures. Chemicals that are dissolved in water are said to be in the *aqueous* phase. Solvents other than water are known as *cosolvents*, the most important of which is petroleum. A solute can have very different solubility when a cosolvent is present. There are numerous chemicals that are more soluble in petroleum than in water, which has important implications for their transport in the subsurface.

When the chemical released is in its pure rather than dissolved phase, it is known as a *nonaqueous phase liquid*, or NAPL. There are two types of NAPL, each defined by the chemical's *specific gravity*, which is the density as compared to water. A compound with a specific gravity less than 1.00 is lighter than water. The portion of that compound that does not dissolve in water and is in the liquid phase is called a *light nonaqueous phase liquid*, or LNAPL. LNAPL will float on top of the water, be it above-ground surface water or below-surface groundwater. A compound with a specific gravity greater than 1.00, known as *dense nonaqueous phase liquid*, or DNAPL, will sink through the water column, with important implications to environmental investigations and remediations.

4.3.2 The Hydrogeologic Cycle

Water finds its way into the subsurface as part of the *hydrogeologic cycle*. A schematic diagram of the hydrogeologic cycle is shown in Figure 4.4. In the hydrogeologic cycle, water gets deposited on the earth's surface as precipitation or condensation, evaporating or transpiring (which is similar to evaporation, but from plants) back into the atmosphere only to fall back to earth as precipitation or condensation, starting the whole cycle over again. When it falls to earth, the water either becomes part of a surface water body, such as an ocean or a lake, or it lands on an unsaturated surface and infiltrates down into the earth. The water that does not return to the atmosphere will keep working its way through the soils, first

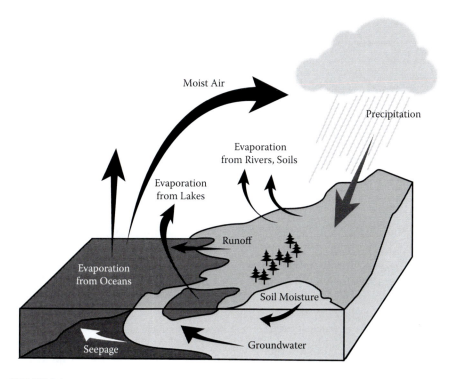

FIGURE 4.4
The hydrogeologic cycle.

manifesting itself as soil moisture in the unsaturated zone, and eventually becoming groundwater in the saturated zone. The groundwater eventually discharges to a surface water body, again becoming a part of the never-ending hydrogeologic cycle.

4.3.3 Vadose Zone

The unsaturated portion of the subsurface that the water moves through is known as the *vadose zone*. The vadose zone is a complicated area where a myriad of forces affect the fate and transport of a chemical.

Biological forces in the subsurface include microorganisms living in the soil, which can ingest the chemical and process it as food. This process can result in *biodegradation*, which changes the chemical into one or more different chemicals. Plant roots present in the first few feet of soil may capture the chemical and take it up through its root system as part of its water uptake.

Various chemical and physical processes will also be at work in the subsurface. Carbon tends to act as a sponge, trapping and retaining many chemicals (this property of carbon comes in handy during remediation, as discussed in Chapter 8). Other chemicals present in the soils, or in gaseous phase in the voids between the soil grains, may react with and alter the chemical. The chemical also may volatilize by spontaneously going into the vapor phase, especially if the chemical is a volatile organic compound (see Chapter 3). Once in the gaseous phase, the chemical may move upward through the vadose zone, possibly creating indoor air problems through a process known as vapor intrusion (see Chapter 7).

If the chemical in the vadose zone remains in the liquid phase (in either a pure or dissolved phase), it will behave like a liquid in the subsurface. It therefore will obey two basic principles that govern fluid flow:

- Liquids always flow downhill, unless under pressure.
- Liquids tend to flow in the path of least resistance.

The chemical generally will move downward until it encounters what is known as the *capillary fringe* (see Figure 4.5). Situated at the boundary of the two zones, the capillary fringe is a portion of the soils, technically in the vadose zone, in which a liquid can rise above the saturated zone due to surface tension. Surface tension is the same process that allows liquids to rise above the rim of a cup without spilling.

Once a liquid reaches the saturated zone, it merges with the water already present in the saturated zone. Water that is present in the saturated zone is known as *groundwater* (sometimes spelled as two words: ground water). Once part of the groundwater, the chemical is subject to a new and often more complex transport regime.

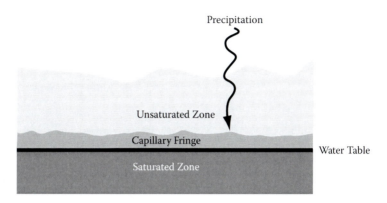

FIGURE 4.5
The capillary fringe is located between the vadose zone and the zone of saturation.

4.3.4 Saturated Zone

In ideal conditions, the top of the saturated zone is a surface whose shape is a subdued reflection of the overlying topography. It mimics the overlying surface in ideal conditions because if the soils in the vadose zone are homogeneous, that is, the same throughout, then rain water and other liquids should percolate downward in a uniform manner, saturating the underlying soils in a uniform manner. The *water table*, which is the interface between the saturated and unsaturated portions of the soils, is aptly named, for diagrammatically (and under ideal conditions) it has the configuration of a tilted table.

The capillary fringe should not be confused with the *smear zone*, which is an area that is saturated on a seasonal basis. In the temperate and moist areas of the United States, the elevation of the water table can vary several feet, creating a portion of the soils that at times is part of the vadose zone and other times lies below the water table.

4.3.4.1 Hydraulic Gradient

When the water table is tilted, the water has a potential energy that becomes kinetic energy as it flows downward. That potential energy is known as the *hydraulic head*. *Darcy's law* expresses in mathematical terms the relationship between the factors that govern the flow of groundwater. In Darcy's equation, the velocity of the water depends on two parameters, as follows:

$$V = ki,$$

where V is the specific discharge (also known as the flow rate, or Darcy velocity), i is the hydraulic gradient, and k is the hydraulic conductivity.

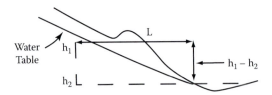

FIGURE 4.6
Calculating hydraulic gradient. (Courtesy of Professor Steven A. Nelson, Tulane University.)

Hydraulic gradient is essentially the slope of the water table. It is the rate of change in which energy, or hydraulic head, is lost as water flows downward through the porous materials. The steeper the water table, the greater the hydraulic gradient. Figure 4.6 shows that the hydraulic gradient, i, is calculated by the slope of a line in a right triangle, as follows:

$$i = \Delta h/l,$$

where l is the flow distance. "Up gradient" refers to a physical location that has a greater hydraulic head than the subject location, while "down gradient" refers to a physical location that has a lesser hydraulic head than the subject locations. A location with the same hydraulic head as the subject location is designated as "cross-gradient."

Hydraulic conductivity is the ability of a formation to transmit water. It is not dissimilar in concept to the more familiar electric conductivity, which is the ability of a material to transmit an electric current. The greater the hydraulic gradient or the hydraulic conductivity, the greater the flow rate of the liquid in the formation.

4.3.4.2 Groundwater Flow

In unconsolidated soils, groundwater, because it flows downhill, generally follows topography. It should be noted, however, that if surface topography is altered, the alterations do not necessarily affect groundwater flow direction. Therefore, especially in flat, gently sloping, or developed areas, the topographic slope should be treated as a first approximation of groundwater flow direction. Confirmation of groundwater flow direction only can be achieved with data collected at monitoring points (see Chapter 6).

Groundwater does not flow downward forever. Eventually, it reaches a point, known as a *groundwater divide*, at which its flow changes direction. In many cases, that divide is indicated by the presence of a water body. Figure 4.7 is a schematic diagram of a *gaining stream*, also known as a *spring-fed stream*. Groundwater on both sides of the stream flows into the stream and provide it with additional water, hence its name. Water flows in opposite directions toward the stream on both sides of the stream, with important

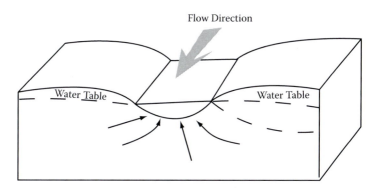

FIGURE 4.7
A schematic diagram of a gaining stream.

implications for contaminant migration (see Chapter 7). The opposite of a gaining stream is a losing stream (see Figure 4.8), common in arid regions, in which some of the water in the stream migrates downward from the stream, seeking a saturated zone.

Water movement in fractured bedrock can be extremely complicated. Figure 4.9 provides an example of the complexities involved in water and chemical movement in bedrock. In this figure, chemicals, in this case a DNAPL, released from a source zone, will travel generally downward through the vadose zone, but through an increasingly set of circuitous pathways. Once the DNAPL encounters the saturated zone, it spreads both downward and down gradient, in the direction of groundwater flow.

Keeping in mind that this figure is in two dimensions rather than three dimensions, fractures that appear to be connected in two dimensions may not actually intersect in three dimensions. This can lead to a migration pattern in which certain fractures become contaminated while fractures above, below, up gradient and down gradient remain untouched. Unless the fracture pattern

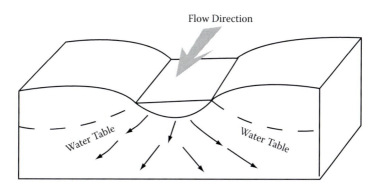

FIGURE 4.8
A schematic diagram of a losing stream.

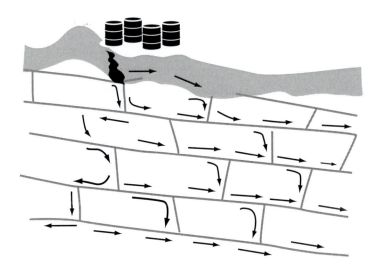

FIGURE 4.9
Complex groundwater flow patterns in fractured bedrock.

can be understood and predicted, the migration of a contaminant in fractured bedrock can be extremely difficult if not impossible to fully understand.

4.3.4.3 Advection

Advection, as used in hydrogeology, is the mechanism by which groundwater and its solutes move down gradient. If advection was the only force affecting water flow in the saturated zone, then it would be fairly easy to understand chemical transport in the saturated zone. Under advective forces, a mass of chemicals released to the subsurface would encounter the water table, then move in the direction of groundwater flow. Such a mass of chemicals is generally known as a contaminant *plume*.

In the real world, chemicals tend not be behave in this manner because of the existence of two other forces: diffusion and dispersion.

4.3.4.4 Diffusion and Dispersion

Diffusion is the movement of a chemical from higher concentration to lower concentration. Diffusion causes chemicals to spread out in directions other than the direction of water flow and can occur in the absence of water flow. As such, it enables a contaminant plume to spread in directions other than the direction of groundwater flow.

A compound's *diffusion coefficient* describes its ability to move from an area of high concentration to an area of low concentration. The amount of compound that will flow through a small area during a small time interval (J, also known as *flux*) is determined using Fick's first law, which states:

$$J = -D(dC/dx)$$

where D is the diffusion coefficient (for a given temperature), C is the concentration of the contaminant in solution, and x is the length in the direction of movement.

Diffusion coefficients are calculated in the laboratory and available in published scientific literature. Therefore, for a given chemical, if the chemical concentration in the center of a chemical plume is known, and the edge of the plume is known, the concentration gradient can be estimated, enabling the geologist to estimate how the plume will spread over a period of time through the process of diffusion.

Except in karst terrains, where water may flow in the subsurface via real underground streams, subsurface water must flow through rather than over geological formations, enabling the geology to impact the flow of the water. Mechanical dispersion, usually known simply as *dispersion*, has the same effect as diffusion in causing chemicals to spread out in directions other than the primary direction of groundwater flow. However, dispersion is a physical rather than a chemical force, caused by the differing fluid velocities within the pores and pathways taken by the fluid.

Figure 4.10 shows the effects of dispersion on groundwater flow. Diagram a shows a liquid moving along a smooth, even pathway through well-sorted sand. Diagram b shows the liquid moving through a tortuous path in an effort to get through poorly sorted sand, that is, one with a wide variety of soil types. The circuitous path translates into slower transport through the

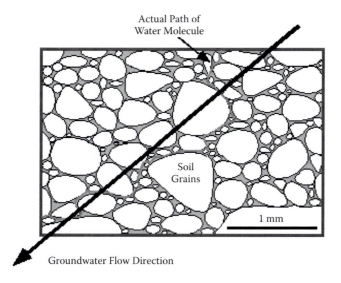

FIGURE 4.10
The effect of grain spacing on the permeability of a soil formation. (From Fetter, C. W., 2000, *Applied Hydrogeology*, 4th ed., New York, Prentice Hall.)

sand and more opportunities to interact with biological agents or chemicals that can impact the fate and transport of the chemical.

Dispersion in various types of soils can be calculated in the laboratory. However, estimating the degree of dispersion in a soil based on laboratory models is a tricky business, and in the field, trickier still. Therefore, estimating dispersion effects is generally done using field data.

4.3.4.5 Retardation and Attenuation

Diffusion and dispersion forces work at preventing dissolved chemicals from moving in the same direction as groundwater flow. There are also forces in the subsurface that work at preventing dissolved chemicals from moving at the same velocity as the groundwater. This slowing of a dissolved chemical is referred to as *retardation*, as opposed to the slowing of the transport of a chemical by reducing the mass of the chemical through biological or chemical processes, which is known as *attenuation*.

Figure 4.10c shows the effect that friction plays on the movement of the liquid. When retardation occurs, the differential velocities of groundwater flow result in a spreading of the chemical. The sum total of the retardation and attenuation of a contaminant plume is known as the *retardation factor*.

One cause of retardation in the transport of a chemical is known as *sorption*, which is the process by which chemicals accumulate on the surface of individual soil grains. The surface that attracts the chemical is known as the sorbent, and the chemical that is attracted to the sorbent is known as the sorbate. The tendency of a chemical to be adsorbed by soils or sediment is known as its *soil–water partition coefficient*.

A particular type of soil–water partition coefficient is the *organic carbon partition coefficient*, which describes the ability of organic carbon present in the otherwise inorganic, mineralic soils to inhibit a chemical's movement. Organic carbon's ability to affect a chemical's mobility plays a role in certain remedial technologies (see Chapter 8), especially those involving granular activated carbon (GAC), an especially effective sorbent.

4.3.4.6 Aquifers

Saturated soil that can transmit a "significant" quantity of liquid is known as an *aquifer*.[*] For an aquifer to exist, the soil must have sufficient porosity to hold an appreciable quantity of water, and sufficient permeability to allow the water to move through the formation. Aquifers can occur in bedrock as well, as long as the bedrock has large, interconnected fractures.

---

[*] This is one of many definitions of what constitutes an aquifer. The environmental consultant should know what the governing regulatory body considers an aquifer before undertaking an environmental investigation within that jurisdiction.

4.3.4.7 Aquitards, Aquicludes, and Confining Layers

Well-sorted sands and gravel formations make for good aquifers. Soil formations that have a low porosity or permeability, and thus inhibit the flow of liquids in the subsurface, are known as *aquitards*. Aquitards are typically composed at least partly of silts or clays. Soil formations that are effectively impermeable are known as *aquicludes*.

Aquifers, unless they are overlain by aquitards or aquicludes, can receive water from the surface that has percolated down through the vadose zone. This process is known as *recharge*. A soil formation that acts as an aquitard or an aquiclude in preventing or inhibiting water from recharging the aquifer is known as a *confining layer*. It should be noted that a confining layer works both ways, also preventing upward movement of groundwater, as discussed later.

Confining layers greatly impact the configuration of the water table. In Figure 4.11, water percolates into an aquifer that is overlain by an aquitard. In this case, the water table would naturally rise above the aquitard were it not in the way. The aquitard therefore depresses the water table, keeping the water in the aquifer under pressure. When the pressure is relieved, as happens when a drinking water well penetrates the aquitard and reaches the aquifer, the groundwater rises to its desired level. The level to which the water rises is known as the *potentiometric surface* (also called the *piezometric surface*), so named because it is the surface at which groundwater reaches its greatest potential energy, or head.

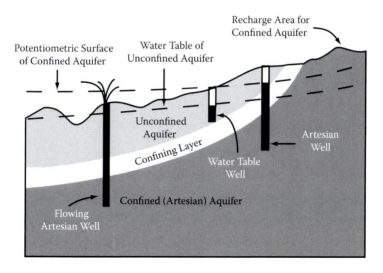

FIGURE 4.11
Diagram showing the existence of flowing artesian well conditions due to the potentiometric surface being at a higher elevation than the land surface. (Adapted from the Minnesota Department of Health.)

As shown in Figure 4.11, when the potentiometric surface of the groundwater is above the ground surface, groundwater will flow upward to its potentiometric surface and onto the ground. This is known as an *artesian condition*. (It should be noted that some people consider an artesian condition to exist in any situation where the water table rises once the confining layer has been breached.) It is important to remember that even under artesian conditions the groundwater is still obeying the two principles of liquid flow, since the groundwater beneath the confining layer was under pressure until the pressure was relieved by the drilling of the well.

4.3.4.8 Perched Aquifers

A perched aquifer is formed when a low-permeability formation, such as a clay layer, is present within a higher permeability formation, such as a sand layer (see Figure 4.12). This causes water percolating downward to become trapped by the low-permeability formation and pool there, resulting in a discontinuous zone of saturation. Scientists debate even the name of this formation, preferring to call it a *perched water table*, since calling it an aquifer implies the potential to transmit a quantity of liquid that the discontinuous zone is unlikely to produce.

It is often difficult to distinguish a perched aquifer from the underlying saturated zone, since both are saturated. In theory, the perched aquifer is discontinuous and underlain by an unsaturated zone while the underlying aquifer is continuous and underlain by additional saturated formations.

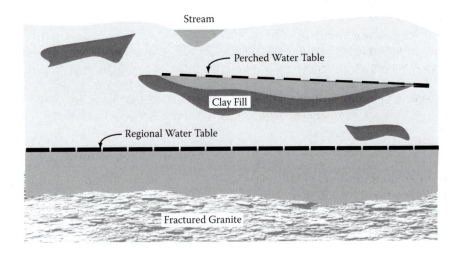

FIGURE 4.12
A perched water table aquifer. Impermeable layers that do not form a bowl to hold water will not form a perched water table aquifer. (Adapted from Sterrett, Robert J., ed., 2007, *Groundwater and Wells*, 3rd ed., Johnson Screens.)

However, in practice a perched aquifer with significant lateral extent can look a lot like a regular aquifer to the field observer.

Perched aquifers are important to subsurface investigations since the low-permeability formation not only stops the downward movement of water but also the downward movement of chemicals that are being transported in the water.

Having established the regulatory and scientific framework under which environmental investigations and remediations are conducted, the next several chapters discuss the methodologies by which they are conducted.

References

ASTM International. 2010. ASTM D2487–10: Standard Practice for Classification of Soils for Engineering Purposes (Unified Soil Classification System).

ASTM International. 2009. ASTM D2488–09a: Standard Practice for Description and Identification of Soils (Visual-Manual Procedure).

Fetter, C. W. 2000. *Applied Hydrogeology*, 4th ed. New York: Prentice Hall.

Freeze, R. Allan, and Cherry, John A. 1979. *Groundwater*. New York: Prentice Hall.

Hemond, Harold F., and Fechner-Levy, Elizabeth J. 2000. *Chemical Fate and Transport in the Environment*, 2nd ed. Salt Lake City, UT: Academic Press.

Lagrega, Michael D., Buckingham, Phillip L., and Evans, Jeffrey E. 2001. *Hazardous Waste Management*, 2nd ed. New York: McGraw-Hill.

Nyer, Evan K., and Gearhart, Mary J. Winter 1997. Plumes don't move. *Ground Water Monitoring and Remediation*, 27(1): 52–55.

Soil Survey Staff. 1999. *Soil Taxonomy: A Basic System of Soil Classification for Making and Interpreting Soil Surveys*. 2nd ed. Washington, DC: U.S. Department of Agriculture Natural Resources Conservation Service.

Sterrett, Robert J., ed. 2007. *Groundwater and Wells*, 3rd ed. Johnson Screens.

5

Phase I Environmental Site Assessments

The Phase I Environmental Site Assessment is the first step for many if not most environmental investigations. There are many varieties of Phase I ESAs and they go by a variety of names, such as preliminary assessments, preliminary site assessments, or Phase I audits. But in most places and to most people, they are simply known as a Phase I.

5.1 History of the Phase I ESA

The main impetus for the creation of the Phase I is the federal Superfund law. As discussed in Chapter 2, the 1986 Superfund Amendments and Reauthorization Act (SARA) provided for an *innocent purchaser defense* from Superfund liability. To qualify for the innocent purchaser defense, the person had to have (1) acquired the property after the pollution had occurred; and (2) had to demonstrate that before the acquisition, it had no knowledge and no reason to know of the contamination. To demonstrate that the person had no reason to know of the contamination, the person was required to have made *"all appropriate inquiries* [italics mine] into the previous ownership and uses of the facility in accordance with generally accepted good commercial and customary standards and practice."

Years went by without word from the U.S. Environmental Protection Agency (USEPA) on what actually comprised "all appropriate inquiries." In the meantime, the influx of environmental regulations at the federal and state levels heightened the sensitivity of the business community to the potentially detrimental effects of environmental liabilities on their business. By the late 1980s, prospective purchasers of commercial properties had begun hiring environmental consultants to perform basic due diligence activities in support of commercial real estate transactions. These activities had much in common with today's Phase I but could not be formally represented as constituting an innocent purchaser defense under Superfund since no such protocol had yet been established.

A not-for-profit organization, then known as the American Society for Testing and Materials (now known as ASTM International, or simply ASTM), stepped into the breach. ASTM, which then as now sets standards for products and services in the United States, formed a committee in March 1990 to develop a standard practice for environmental due diligence in property transactions. That committee issued the first Standard for Environmental Assessments in Real Estate Transactions in 1993, known as E1527. The ASTM standard quickly became the standard for the entire industry. Other major organizations, such as insurance companies, had developed their own standards, but after 1993 virtually all of them were variations on the ASTM standard.

The ASTM standard was revised in 1994, 1997, and 2000 to reflect advances in technology and the evolution of the usage of the Phase I as a tool to evaluate business risk in a property transaction posed by environmental conditions. The next revision came in 2005 in response to the passing in 2004 of the USEPA All Appropriate Inquiries Rule. This rule was the long-awaited delivery on the requirement in the 1986 SARA legislation. ASTM had significant input into the development of the rule, so it was no surprise that the ASTM standard in existence at that time substantially complied with the rule.

One significant change made in the 2005 standard was the expansion of liability protections to two groups other than the innocent purchaser:

Bona fide prospective purchaser—One who buys property after conducting all appropriate inquiry into the current and historical uses of the property knowing that there are environmental conditions present on the property.

Contiguous property owner—One who owns property that is contiguous to, and may be impacted by, hazardous substances migrating from property they do not own. The required qualifications to obtain this liability protection are discussed later in this chapter.

Another major change in the 2005 standard was the creation of the term *Environmental Professional*. The Environmental Professional (EP), according to the ASTM Phase I standard, is a person experienced at conducting Phase I ESAs and has the necessary education and training to be qualified for the job. The EP is responsible for the findings of the ESA. Other people may perform the various activities entailed in an ESA, but the Environmental Professional is the person who is held accountable for the results.

The 2005 standard included the need to investigate for "controlled substances" as part of a Phase I ESA. Controlled substances could be an environmental issue because of the usage of certain polluting chemicals in processes that create methamphetamine and other illegal substances. However, an investigation into the presence of controlled substances on a property must be included in the scope of a Phase I ESA only if the Phase I ESA is funded with a federal grant under the EPA Brownfield Assessment and Characterization Program.

5.2 The Phase I ASTM Standard—Recognized Environmental Conditions and Historical Recognized Environmental Conditions

A Phase I ESA is a procedure that assesses whether a piece of real estate has a potential environmental problem, or what the E1527 standard refers to as a "recognized environmental condition." The standard defines a *recognized environmental condition*, or REC, as follows:

> The presence or likely presence of any hazardous substances or petroleum products on a property under conditions that indicate an existing release, a past release, or a material threat of a release of any hazardous substances or petroleum products into structures on the property or into the ground, ground water, or surface water of the property.

This definition contains three terms that merit discussion.

Hazardous substances or petroleum products—Figure 5.1 shows a vacant lot strewn with debris, while ugly, the mere presence of wastes on the property does not constitute an REC unless there has been a release of hazardous substances or petroleum (as defined in Chapters 2 and 3).

Property—The Phase I ESA only deals with commercial real estate, which is a property that can be bought or sold. It does not apply to mobile sources of pollution, such as vehicles, boats, or airplanes, nor does it apply to properties that are not typically bought or sold,

FIGURE 5.1
The presence of refuse and other inert wastes by themselves does not constitute an REC. (Courtesy of GZA GeoEnvironmental, Inc.)

such as public water bodies, public parks, and public roadways. The ASTM standard excludes residential properties with buildings in which there are less than four dwelling units. The assumption is that if a dwelling has four or more units, its principal function is revenue generation and is therefore commercial real estate.

Material threat of a release—An REC does not require the existence of a release or a past release. The material threat of a release is sufficient to establish an REC. A *material threat* is a physically observable or obvious threat that is reasonably likely to lead to a release that might result in impact to public health or the environment. "Obvious" is the operative term here. If one has to work hard in developing a scenario in which an observed or otherwise identified condition would result in a release to the environment, then it is not a material threat and does not qualify as an REC.

ASTM E1527 goes on to state that an REC

is not intended to include de minimus conditions that generally … would not be the subject of an enforcement action if brought to the attention of appropriate governmental agencies. Often, it is left to the judgment of the inspector whether a spill is of sufficient size to be considered an REC.

Drips of motor oil underneath a car in a parking lot would therefore not be classified as an REC, whereas the more significant petroleum staining shown in Figure 5.2 might be. State regulations also weigh in on what constitutes a reportable spill.

ASTM E1527 also defines a *historical recognized environmental condition*, or HREC. An HREC is defined as an environmental condition that, in the past, would have been considered an REC, but would not currently be considered an REC, usually because the offending circumstance had been alleviated. For instance, an oil spill would be considered an REC, but an oil spill that has already been cleaned up to the satisfaction of the relevant regulatory agency would be considered an HREC.

It is important to note that an HREC is not an REC due to a historic condition. A chemical release that occurred 30 years ago and has not been cleaned up is an REC. Once it is cleaned up, it is no longer a concern to the regulatory community or to the Environmental Professional and can be considered an HREC rather than an REC.

5.2.1 Exclusions from the Standard

Given its origins as a CERCLA defense, Standard Practice E1527 addresses only conditions that have the potential to impact the environment. There are numerous environmental issues that are excluded from this definition. The

FIGURE 5.2
Significant oil staining near an aboveground storage tank. (Courtesy of GZA GeoEnvironmental, Inc.)

E1527 standard lists the following environmental issues as being explicitly excluded from the Phase I ESA:

- Asbestos-containing materials (ACM)
- Radon
- Lead-based paint (LBP)
- Lead in drinking water
- Wetlands
- Regulatory compliance
- Cultural and historic resources
- Industrial hygiene
- Health and safety
- Ecological resources
- Endangered species
- Indoor air quality
- High-voltage power lines

The exclusion of these environmental issues in no way minimizes their potential impact on a property or on human health. These conditions, because they can affect the value of a property without causing any implications regarding CERCLA liability, are lumped together under the term *business environmental risk*. The E1527 standard defines business environmental risk as a risk that can have a material environmental or environmentally driven impact on the current or planned use of a parcel of commercial real estate. Many of the excluded environmental concerns are discussed in subsequent chapters to this book.

5.3 Components of the ASTM Standard

Determining if (and where) pollution has occurred at a site, such as the abandoned factory shown in Figure 5.3, can challenge the most experienced site assessor. The E1527 standard provides a methodology for rooting out most of the salient issues present on such a property.

The research portion of the Standard Practice E1527 consists of six steps:

1. Site and vicinity reconnaissance
2. Site interviews
3. User responsibilities
4. Site history review
5. Local agency review
6. Regulatory review

Each of these steps is described next.

FIGURE 5.3
An abandoned factory. (Courtesy of GZA GeoEnvironmental, Inc.)

5.3.1 The Site and Vicinity Reconnaissance

The site and vicinity reconnaissance is the most obvious and visible portion of the Phase I ESA. The basic building blocks of the site and vicinity reconnaissance are areas of concern (AOCs). AOCs can be defined as areas where hazardous substances or petroleum products could be located or could have been located in the past. Table 5.1 lists AOCs that can be encountered at a site. Please note that the mere presence of any of the AOCs listed in Table 5.1 does not constitute an REC. The Environmental Professional should assess each of these AOCs to evaluate whether their presence is consistent with the criteria dictating an REC.

The most obvious RECs are visually observable releases. The inspector will look for staining on paved and unpaved surfaces. It is important to be able to distinguish between staining caused by hazardous substances and petroleum products from staining caused by water and other substances not of concern. A rainbow sheen typical of petroleum floating on water is also direct evidence of a release as are chemical odors in the ambient air.

The presence of dead or sickly looking vegetation, as observed around the oil spill shown on Figure 5.4, can be indicative of a spill of petroleum or hazardous substances. The inspector must judge whether the stressed vegetation is due to a spill or to other reasons, such as lack of water, lack of sunlight, or foot traffic. Although the spilled chemicals may have disappeared, the lingering effects on the flora in the area can be direct evidence of contamination.

Any container whose contents are hazardous substances or petroleum product has the potential to leak, spill, or otherwise empty its contents into the environment. However, containers by themselves are not RECs, otherwise every hardware store, auto parts store, and home improvement center would be filled with RECs. The other crucial element to defining an REC is a pathway by which the contents of a vessel container could reasonably enter the environment.

The following sections discuss some of the more important and more common potential environmental concerns, and some of the evidence to be sought by the investigator in deciding whether an AOC is in fact an REC.

5.3.1.1 Underground Storage Tanks and Filling Stations

Although any vessel that stores or has stored hazardous substances or petroleum merits a concern, certain vessels by merit of their size or their location are of greater concern than others. Of very high concern are *underground storage tanks* (USTs), which is a bulk storage vessel with at least 10% of its exterior in contact with the ground. USTs can contain hazardous substances or petroleum. However, the overwhelming number of USTs contains petroleum, usually fuels for motor vehicles or heating oil for heating buildings.

Because the UST is in direct contact with the ground, any leak from that portion in contact with the ground will result in a release into the

TABLE 5.1

Typical Areas of Concern

Bulk Storage Tank Systems Areas of Concern
Aboveground storage tanks and associated piping
Underground storage tanks and associated piping
Silos
Rail cars
Loading and unloading areas
Piping, above ground and below ground pumping stations, sumps, and pits

Storage and Staging Areas of Concern
Storage pads including drum and/or waste storage
Surface impoundments and lagoons
Dumpsters
Hazardous material storage or handling areas

Areas of Concern Regarding Drainage Systems
Floor drains, trenches, piping, sumps
Process area sinks and piping that receive process waste
Roof leaders when process operations vent to the roof
Drainage swales and culverts
Storm sewer collection systems
Storm water detention ponds and fire ponds
Surface water bodies

Areas of Concern Regarding Discharge and Disposal Areas
Septic systems, seepage pits, leach fields, sumps, and dry wells
Landfills or landfarms
Sprayfields
Incinerators
Historic fill or any other fill material
Open pipe discharges
Noncontact cooling water discharges
Areas that receive flood or storm water from potentially contaminated areas
Active or inactive production wells

Building Interior Areas of Concern
Loading or transfer areas
Waste treatment areas
Boiler rooms
Air vents and ducts
Hydraulic elevators

Other Areas of Concern
Electrical transformers and capacitors
Waste treatment areas
Discolored or spill areas

TABLE 5.1 (CONTINUED)

Typical Areas of Concern

Areas of stressed vegetation
Underground piping including industrial process sewers
Compressor vent discharges

Source: Adapted from New Jersey DEP.

FIGURE 5.4
Dead, brown vegetation borders this oil spill in Bronx, New York. (Courtesy of GZA GeoEnvironmental, Inc.)

environment. USTs also store large quantities of liquids, typically 500 gallons up to 20,000 gallons or more. There are hundreds of thousands of USTs in the United States, making USTs one of the most common sources of RECs in Phase I ESAs.

One place where USTs are guaranteed to exist is a filling station. Most filling stations have multiple USTs containing various grades of gasoline. Visual evidence of their existence is limited to a fill port, which is used to deliver the liquid to the UST for storage and eventual usage, and a vent pipe, which enables air to escape from a UST to make room for the liquid when the UST is being filled.

As shown in Figure 5.5, USTs are not the only areas of concern at a filling station. There is also underground piping that allows the petroleum product to flow into the dispensers. The joints and elbows in the underground piping are particularly vulnerable to leakage. The dispensers themselves are areas of concern, especially due to the possibility of spills from the filling of vehicular gas tanks. Filling stations that are also vehicle service stations may contain motor oil USTs and waste oil USTs that are often buried beneath the surface due to limited space in the station.

Most USTs, however, store heating oil. They can be found at single-family residences, multitenant apartment buildings, and commercial and industrial facilities—in other words, in just about any building that is heated.

Even more insidious than USTs are former USTs. Natural gas lines from the oil patch to the northeast and midwest United States were constructed in the 1960s and 1970s, enabling many buildings to convert their heating

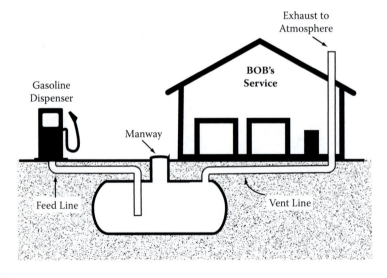

FIGURE 5.5
Conceptual diagram of a gasoline filling station. The manway provides physical access to the people responsible for maintaining or repairing the UST system.

systems from oil to natural gas. A building constructed prior to 1960 in these areas currently may have natural gas as its heating fuel source. However, it probably was not constructed that way—there may be a relict oil heating system still on the property, including an abandoned heating oil UST.

There may be no visual evidence of the existence of former USTs. Sometimes the former fill port or vent pipe still exists. A large patch in the asphalt or concrete outside of a building hints at the possibility that a UST was removed from that spot (although it may just hint at the possibility that the area needs to be repaved). The presence of natural gas heating in a building that is 1950s vintage or older begs the question of how the building was heated before natural gas became widely available to the area, if such is the case.

5.3.1.2 Aboveground Storage Tanks

An *aboveground storage tank* (AST) is a storage vessel in which less than 10% of the tank is in contact with the ground. ASTs can be present in the basement of a building, even though the tank is below the ground level.

Despite having similar capacities and holding similar substances, ASTs generally do not represent the same hazard to the environment as USTs because, in most cases, they are not in direct contact with the ground.* Instead, they often are suspended above the underlying surface by steel legs or concrete cradles (see Figure 5.6). Most releases from an AST not only do not have a direct pathway to the environment, but often the integrity of ASTs is relatively easy to verify. Therefore, damaged ASTs are more likely to be fixed or replaced than their underground cousins before they can impact the environment.

5.3.1.3 Drum Storage Areas

Smaller bulk storage containers, such as 55-gallon drums, pose a similar threat to the environment as an AST. Though they store a smaller volume of hazardous substances or petroleum than an AST, they are typically present in groups and therefore pose, as a group, a major concern if in damaged condition or stored improperly. The Environmental Professional must judge whether a combination of a drum's ability to release a material quantity of its contents and the existence of a reasonable pathway to the environment constitutes an REC.

---

* The regulatory definition of an underground storage tank is a storage tank in which no more than 10% of the tank is in contact with the ground. Oil terminals and other bulk storage facilities tend to have ASTs, often with a capacity of 50,000 gallons or more, whose bottoms are in contact with the ground. Because of their large storage capacity, releases from these ASTs can cause widespread contamination. Most ASTs, however, tend to be far smaller (1000 gallon capacity or less) and are not in contact with the ground, thereby posing a far lesser threat to the environment than USTs.

FIGURE 5.6
Aboveground storage tank within secondary containment. (Courtesy of GZA GeoEnvironmental, Inc.)

5.3.1.4 Industrial Establishments

Industrial establishments that store and handle chemicals or petroleum products are high on the list of concerns for the EP. One way to conceptualize an industrial operation is a three-step process regarding hazardous substances and petroleum: materials and chemicals come into the facility, materials and chemicals get processed at the facility, and products and wastes leave the facility. As a general rule, the places where hazardous substances and petroleum are most often handled are the places where spills are most likely to occur. These places include loading and unloading areas; areas where chemicals are introduced into a process or into machinery; and places where wastes are containerized for eventual disposal (see Figure 5.7). Pathways for wastes to escape the process area—floor drains, floor trenches, wastewater lines—are of particular interest to the inspector.

Typical areas where hazardous substances or petroleum are stored or used include boiler rooms, mechanical rooms, maintenance and janitorial rooms, closets, and basements. In a manufacturing facility, the inspector must develop a basic understanding of the processes occurring to understand where the hazardous substances or petroleum products and wastes are located, and where the potential migration paths to the environment are located so that AOCs can be evaluated. Soliciting the input of someone knowledgeable of the facility's processes, such as the plant manager or the facility engineer, is crucial to understanding the risks associated with plant operations and is required under the E1527 standard.

An industrial facility may also have smaller containers located at various portions of the facility. Machinery might also contain reservoirs of

FIGURE 5.7
Fifty-five-gallon steel drums stored on their side near a storm drain.

petroleum, such as hydraulic equipment (including elevators and lifts) and air compressors. Each area where hazardous substances are stored or used is considered to be an AOC, and the Environmental Professional must judge whether the AOCs are RECs.

Industrial facilities often keep detailed records of chemical usage and chemical waste disposal. Such records may be in the form of regulatory filings under various statutes, including hazardous waste disposal manifests under the Resource Conservation and Recovery Act (RCRA), and toxic release inventories, community right-to-know filings, and material safety data (MSD) sheets. Sometimes relevant records can be located in the engineering department, the maintenance department, or even the purchasing department. It is up to the Environmental Professional to ask the appropriate questions to the appropriate people at the facility to root out the information needed to complete the Phase I ESA.

5.3.1.5 Dry Cleaners

Dry-cleaning establishments are worthy of special mention. Ubiquitous along commercial roads and in shopping centers, the USEPA estimates that there are more than 25,000 facilities in this country where dry cleaning is done on the premises. Dry cleaners, along with gasoline filling stations and auto repair facilities (and printers before the advent of desktop publishing), are the most common users of chemicals that can contaminate the subsurface.

Whereas early dry-cleaning operations used kerosene, a petroleum distillate, and later carbon tetrachloride, a highly toxic petroleum-based solvent, since the mid-1930s the vast majority of dry cleaners have used perchloroethene, known in the industry as perc. The physical and chemical properties of perc are discussed in Chapter 3.

Releases from dry cleaners can occur due to incidental spillage from unloading perc from outside into the machinery; removing waste perc from the machinery and loading it into a waste hauler; handling waste machine filters; and from condensation of vapors that build up inside the dry cleaners, which, once dissolved, can move rapidly through the subsurface. Leakage from improperly maintained machines or from underground piping also can be a major source of perc contamination. In summary, an improperly operated dry cleaners is neither "clean" nor "dry!" Due to its physical properties, it does not take a major spill of perc to cause a major problem.

5.3.1.6 Septic Systems

An AOC that warrants close attention is the septic system. Septic systems are present where no municipal sewer system is available to handle sanitary wastes. They are common in rural areas and in some suburban communities. A problem arises if wastewater containing petroleum or hazardous wastes is being or has been introduced into the septic system.

A typical septic system, shown in Figure 5.8, is nothing more than a waste pipe leading to an underground holding tank, usually made of concrete. Solids settle out in this tank and are periodically pumped out by a septic tank cleaning service. The liquids are either pumped out by the cleaning service or dispersed into the ground through a series of perforated pipes known as *laterals*. When petroleum or hazardous wastes are introduced into a septic system with laterals, the wastes are able to pass through the laterals directly into the soil, with a great chance of then spreading throughout the area.

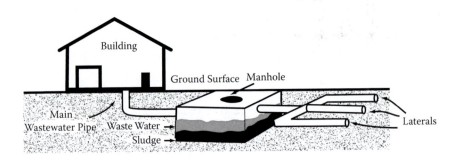

FIGURE 5.8
A typical septic system. The manhole is the access point used to clean out the septic tank.

5.3.1.7 Electric Transformers

High-voltage electrical current available from power lines must be converted to lower-voltage electric current for it to be usable for common purposes. Larger industrial and commercial facilities typically have their own "step-down" transformers to perform the energy conversion. Because of the heat generated during the conversion process, step-down electrical transformers, such as the pad-mounted transformer shown on Figure 5.9, may use dielectric fluid to cool their internal equipment. Dielectric fluids are petroleum based and often contained PCBs prior to 1979 (see Chapter 3). Transformers in which PCB-containing fluids are present still exist. Those that have been retrofitted with non-PCB-containing fluids still may have discharged PCBs to the ground prior to being retrofitted.

5.3.1.8 Evidence of Subsurface Investigations

Subsurface investigations are conducted at sites where contamination is suspected. To investigate groundwater, monitoring wells are typically installed. The presence of a monitoring well at a site should be a red flag that subsurface contamination may exist or, at least, that subsurface contamination has been suspected at the site in the past. The presence of multiple monitoring wells strongly implies that the first monitoring well installed in that area encountered contamination and that the other monitoring wells were installed to delineate the extent of that contamination. Other evidence of a subsurface investigation, such as patched boreholes, also alerts the EP to a potential issue on a property.

FIGURE 5.9
A pad-mounted electrical transformer such as the one shown may use dielectric fluid to cool the internal equipment. (Courtesy of GZA GeoEnvironmental, Inc.)

5.3.1.9 Adjoining Properties

The E1527 standard identifies an *adjoining property* as any real property or properties with contiguous or partially contiguous borders with the site. However, the definition also includes any real property or properties that would have contiguous or partially contiguous borders if not for the presence of a public thoroughfare or other noncommercial real estate. For instance, on Figure 5.10, the property at 100 Elm Street adjoins 104 Elm Street. It also adjoins 101 Elm Street, because if Elm Street, a public thoroughfare, was not present, then 100 and 101 Elm would share a border. On the other hand, 100 Elm Street does not adjoin 105 Elm Street or 97 Elm Street, because, if Elm Street and First Avenue were not present, these properties would share a corner but not a border.

Adjoining properties have special significance because of their ability to impact the subject property. Spills near or on a property can easily migrate onto the next property—pollutants have no knowledge of property boundaries. This is especially true if there is a surface water body, such as a river, available to transport contaminants, or if the contaminants have reached groundwater (more about groundwater later).

If contamination originates on an adjoining property, then the owner of the adjoining property is responsible for cleaning the contamination, unless it can qualify as a contiguous property owner under the All Appropriate Inquiry (AAI) rule. To qualify for this exemption, the property owner must demonstrate through all appropriate inquiry that they did not contribute to the contamination. If the property owner does not qualify for this protection, it may be necessary to perform an extensive investigation involving the installation of boreholes or monitoring wells, the collection of groundwater or soil samples, and expensive analyses of those samples. Even then, the results may be ambiguous, especially if the pollutants in question were also used and possibly spilled on the subject property.

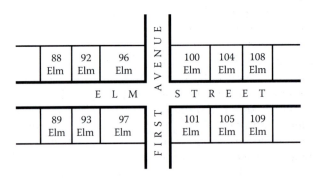

FIGURE 5.10
A schematic diagram of a typical set of properties.

5.3.1.10 Site Vicinity

Although not as significant as adjoining properties, nearby facilities also can impact a site via the migration of contaminants in surface waters or groundwater. The evaluation of nearby properties with the potential to impact the site is discussed in Section 6.3.6.

5.3.2 Site Interviews

Of no less importance than the site reconnaissance are the site interviews. The following are the goals of the site interviews:

- Identify past uses of the site
- Identify specific chemicals present or once present on the site
- Knowledge of spills or other chemical releases
- Knowledge of past environmental cleanups on the site

The E1527 standard requires the inspector to interview three classes of people: the key site manager, major occupants, and persons with actual knowledge. The key site manager, as the title suggests, is the person with the most knowledge about the site, about the facility, and about the facility's current and past operations. If the key site manager is relatively new to the site, then it is the responsibility of the Environmental Professional to identify and interview the former key site manager, or someone else at the facility who has comprehensive knowledge of the four issues noted earlier.

Interviewing major occupants is crucial at facilities where there are multiple occupants, such as shopping centers or multitenant office buildings. These occupants have more significant control of the facility and its operations, and tend to have more history at the facility than smaller tenants. It is important to interview smaller tenants as well, especially if their operations involve the handling or storage of hazardous substances or petroleum, but the E1527 standard specifies only the major occupants to avoid placing an undue burden on the Environmental Professional assessing facilities with numerous tenants.

Identifying persons with actual knowledge is indispensable to the Environmental Professional trying to piece together the history of the property. Actual knowledge is what we might call first-hand knowledge, in that the person does not rely on records or someone else's testimony or memory, but rather on his or her own. Although people's memories can often be false or misleading, it is an excellent opportunity for the Environmental Professional to obtain information not available in the written records. Typical people with actual knowledge are longtime employees or residents and facility operators. The employees need not be upper or even middle management types. In fact, the people "down in the trenches" often have more actual knowledge of "what really went on" than the executive in the corner office.

The E1527 standard requires an interview with at least one contiguous property owner or occupant if the site is abandoned property or in general disrepair, or if there is evidence of "unauthorized uses or uncontrolled access" to the property. This requirement foresees that the owner of an unused property may not be as aware of events occurring on the property as the next-door neighbor, who may be more likely to witness these events. The standard does not specify whether more than one adjoining property owner needs to be interviewed—this is left to the judgment of the Environmental Professional. Certainly, if the adjoining property owner interviewed provides little or no knowledge of the property of interest, the inspector should move on to the next property, hoping for better luck.

5.3.3 User Responsibilities

The 2005 revision to the E1527 standard added a new step to the Phase I ESA. Known as "User Responsibilities," it is an acknowledgement of a simple fact—the person considering the purchase or financing of a piece of real estate has information at his or her disposal that may have bearing on the Phase I ESA.

A typical purchase or financing of a piece of real estate involves many professionals other than the Environmental Professional. Consulting engineers, surveyors, title search professionals, and others generate reports, maps, and so forth in support of the pending deal. Any of these people may come across information that can assist the Environmental Professional in the identification of RECs on the site.

The user, as defined under the E1527 standard, is the person who will use the Phase I ESA report in making a business decision. In general, it is the person who retains the Environmental Professional to perform the Phase I ESA, although sometimes there is an intermediary retaining the Environmental Professional, such as an attorney.

The E1527 standard requires the user to divulge information on at least three topics:

Presence of environmental cleanup liens on the property—A lien is a legal claim against an asset that must be paid when the asset is sold. An environmental cleanup lien is typically placed on a property by a federal, state, or local regulatory authority when the property owner has been negligent in performing a mandated cleanup. With the lien in place, the regulatory authority guarantees that the property will not be sold unless the lien is satisfied at the time of sale. The implications for the presence of RECs on such a property are obvious.

Consideration of "specialized knowledge"—In this context, "specialized knowledge" is knowledge either supplied to or generated by the user that may have environmental implications. For instance, the surveyor or property assessor may encounter a vent pipe, fill port, or some other manifestation of a UST that is missed by the Environmental Professional. Or perhaps the title search professional comes across a document describing the former site usage as a dry cleaner. Often, confidential information is exchanged in the course of a property transaction or financing that the Environmental Professional may not know exists. The E1527 standard requires the user to provide any specialized knowledge that he or she may have about a site to the Environmental Professional.

Relationship of purchase price to fair market value of property, if not contaminated—As the old adage goes, if something is too good to be true, it usually is. If a property is being sold at a price well below market value, there may be a good reason. In particular, the property owner may have been compelled to discount the price because of the presence of known or perceived contamination on the property.

5.3.4 Site History Review

Most RECs, indeed, most contamination, occurred in the past, leaving no visible evidence in the present. The E1527 standard requires the Environmental Professional to identify the "history of the previous uses of the property and surrounding area" to identify past uses that have created RECs. Discovering RECs from historical practices or past events requires the ability to seek out and sift through clues, interpret them, and postulate on the potential occurrence of spills and releases in the past on the site and on surrounding sites. This part of the Phase I ESA more than any other part requires the Environmental Professional to be a detective.

The E1527 standard requires the Environmental Professional to construct a site history back to 1940 or the site's first development, whichever occurred first, with no less than 5-year intervals separating historical information sources.

Evidence of historical spill events or releases can come from numerous sources. Four sources in particular, however, yield the most information to help the Environmental Professional piece together the history of the site and the surrounding area. They are:

- Historical aerial photographs
- Fire insurance maps
- Street directories (also known as city directories)
- Local agency records (which also yield information about current operations, and are discussed in a separate section later)

The first three sources of historical data are discussed next.

5.3.4.1 Historical Aerial Photographs

Aerial photographs were and are commissioned by federal, state, and local governments for planning purposes; by private sector companies to sell to public and private entities for their usage; and by large enterprises for documentation and planning purposes. Dating back to the 1920s, historical aerial photographs are readily available through a number of commercial sources. They can often be accessed for free at public agencies such as county departments of planning, engineering, or transportation.

Aerial photographs provide information on site conditions at the time the photograph was taken. A series of such photos can show the change in property usage over time. Often they are the only evidence for landfilling operations, dumping, or other activities that can contaminate a property that are not related to buildings or industrial processes, especially in rural areas.

For instance, Figure 5.11 is an aerial photograph taken in 1969 of a heavily industrialized area along the Allegheny River in Pittsburgh, Pennsylvania. Figure 5.12 is an aerial photograph of the same properties in 2005, which then (as now) contain a major league baseball stadium.

Aerial photographs vary widely in quality. The best aerial photographs are clear and have good resolution. However, even the best available aerial photographs typically do not offer a scale better than 1″ = 200′. It can be very difficult to discern features on an aerial photograph that are smaller than one acre in size (an acre is 43,560 square feet, or slightly larger than 1′ × 1′ on a high good scale aerial photograph). Depending on the scale of the photograph, buildings may look like dots, so their usage cannot be easily interpreted. However, their presence or absence can be obvious. Integrating aerial photographic interpretations with other data sources can enhance the piecing together of the historic record.

5.3.4.2 Fire Insurance Maps

Fire insurance maps became popular in the United States in the mid-1800s as a way to identify and control risk for the insurance industry. At the time, most structures were constructed with wood, fireproof construction was in its infancy, and the risk of fire to a structure from within or from its neighbors was very real. A cottage industry arose in which individuals physically visited and inspected properties in urban areas and mapped the properties, identifying their construction and any materials that could cause or contribute to a fire. Fortunately for the Environmental Professional, these are the same materials that cause most of the RECs at a site, aligning the professional's interest with that of the insurers from days long past.

Although many companies were active in the fire insurance map business, one company, the Sanborn Map Company, came to dominate the business in the United States by the early 1900s. Therefore, the terms "Sanborn map" and "fire insurance map" are often incorrectly used interchangeably.

FIGURE 5.11
Industrial land usage as seen in 1969. (From GeoPak Historical Aerials, www.historicalaerials.com.)

FIGURE 5.12
Land usage as compared to Figure 5.11 in a 2005 aerial photograph. (From GeoPak Historical Aerials, www.historicalaerials.com.)

The 1950 Sanborn map shown in Figure 5.13 depicts a gasoline filling station with eight circles, which symbolize underground storage tanks. The 1980 Sanborn map of the same location shown in Figure 5.14 shows the western portion of the Brookdale Campus of Hunter College. Other details, such as room usage (swimming pool, amphitheater) are clearly annotated on the map. The map also contains an elaborate system of shorthand notes and symbols. For instance, the number of stories of that part of the building is indicated in Figure 5.13 (2B indicates a two-story building). A key containing an explanation of the shorthand and symbols most often shown on the maps is available from The Sanborn Library, LLC.

5.3.4.3 Local Street Directories

Local street directories were prepared by governmental agencies and private enterprises from the 1800s into the 1960s, when they were supplanted by telephone directories. Rather than listing the residents and businesses in a town alphabetically, they listed them by street address, making them handy references for historical uses of a property and the neighboring area. They are often available in libraries, historical societies, and local governmental offices.

Commercial database providers have geocoded the directories and synthesized the information contained in them to derive city directory abstracts. These abstracts take the data provided in multiple directories and arrange them first by year, then address, then resident/business. Following is a portion of the city directory abstract of a street in Brooklyn, New York (adapted from Environmental Data Resources):

Year	Uses
1928	LIBERTY AVE
	DWOSKIN H GAS STA (885)
	RAMPIANO N & SON ART GLASS (888)
	NACHATOVITZ J GRCRY (890)
	ZALIN WM STA TNY (890)
	LA BARBERA F GRO (892)
	LESHINSKY LOUIS TAILOR (896)
1934	DWOSKIN HYMAN BARBER (881)
	TURFEO JAS TISH DUR H (889)
	TURFEO MARE CLK R (889)
	TORFEO PETER FETYWKR R (889)
	TORFEO RENE FETYWKR R (889)
	MAURO SALVATORE LAB H (890)
	NACHATOVITZ JACOB RETAIL GRO HDO (890)

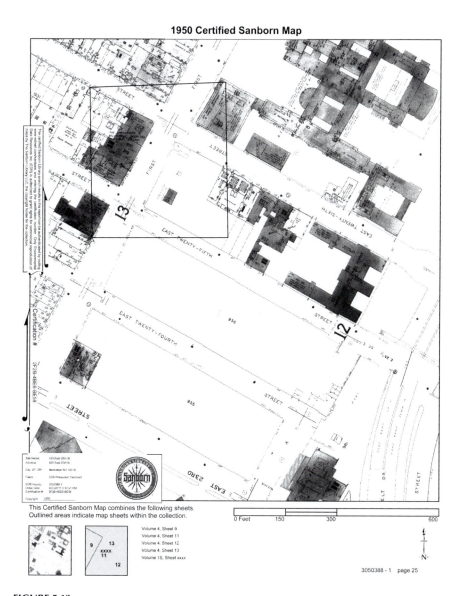

1950 Certified Sanborn Map

This Certified Sanborn Map combines the following sheets.
Outlined areas indicate map sheets within the collection.

Volume 4, Sheet 9
Volume 4, Sheet 11
Volume 4, Sheet 12
Volume 4, Sheet 13
Volume 1S, Sheet xxxx

0 Feet   150   300   600

3050388 - 1   page 25

FIGURE 5.13
Sanborn map of First Avenue and East 25th Street in Manhattan in 1950. Compare land usage to that in Figure 5.14. (From The Sanborn Library, LLC. With permission.)

1980 Certified Sanborn Map

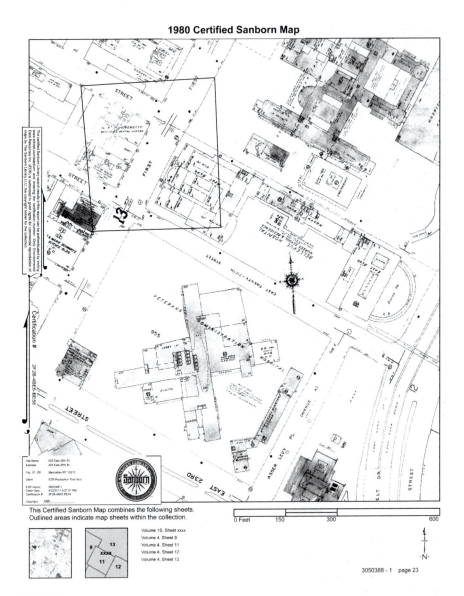

This Certified Sanborn Map combines the following sheets.
Outlined areas indicate map sheets within the collection.

Volume 1S, Sheet xxxx
Volume 4, Sheet 9
Volume 4, Sheet 11
Volume 4, Sheet 12
Volume 4, Sheet 13

3050388 - 1   page 23

FIGURE 5.14

Sanborn map of First Avenue and East 25th Street in Manhattan in 1980. Compared to the land usage in 1950 (see Figure 5.13), no visual evidence remains of the 1950 filling station. (From The Sanborn Library, LLC. With permission.)

This portion of the abstract lists the property at 885 Liberty Avenue as a gasoline filling station in 1928 (some of the abbreviations used in the abstract can be challenging). The 1934 city directory did not have a listing for this gasoline filling station, implying that it was no longer in operation. This gasoline filling station does not appear on any other databases accessed for this property. Without the information provided in the city directory, an important piece of historical information would be lost.

5.3.5 Local Agency Review

Public agencies are often critical sources of information about a property. Agencies at the local level (city, township, or county) are often more readily accessible than state or federal agencies and are more likely to have documents relating to a given property (unless the property is state or federally owned, or otherwise of interest to the state or federal government).

Tax assessor—The tax assessor offers basic information about a property: its size, its ownership, its legal description, and so forth. It is a good starting point for the person who is investigating local records, since this information, especially the legal description of the property, is often the way in which records in other departments are organized.

Registrar of deeds—This department contains historical records of property ownership. Information about past owners of the property may provide clues as to industrial usages of the property or other usages that may have been the source of contamination. One must be cautious in inferring that ownership by an industrial company implies industrial usage. An industrial firm may have had only an office on the property, or may not have utilized the property at all. Conversely, a property that had industrial usage may have been owned by a real estate company, a private individual, or an entity related to the industrial company but whose name does not reveal the connection.

Property deeds themselves can contain clues regarding environmental conditions on the property. They may indicate the presence of landfills on the property, or other practices that would suggest the presence of RECs on the property. The inclusion or exclusion of USTs as part of the sale could be indicated on a property deed, as could language regarding USTs and other environmental conditions, such as the excerpt below from a deed for a property in Greensboro, North Carolina:

9. Install asbestos cap sheet on roof.
10. Install gutter and down spouts where necessary.
11. Excavate hole of sufficient size outside building, at location to be selected by Lessee, for two 10,000-gallon tanks, tanks

to be furnished by Lessee and placed in hole by Lessors in manner to comply with requirements of Building and Fire regulations of the City of Greensboro.

12. Install oil-fired heating system, calculated to maintain a 65 deg. temperature inside building; oil tank (2,000 gal.) to be furnished by Lessee; excavation for burying tank to be done by Lessors; install new door inside furnace room.

Zoning board—Information on the current and former zoning of a property can be obtained at the zoning board. Information on current zoning (e.g., industrial, commercial, residential, or a mixture thereof) can usually be obtained from the tax assessor (see earlier). Although a property that is zoned industrial is not always used for industrial purposes, one can assume that a property that is not zoned industrial is not currently being used for industrial purposes.

Building department—The local building department is often the best source for property-specific information at the town, city, or county level. Because most jurisdictions require the issuance of permits before building construction, renovation, or demolition can be performed, building department records often contain information regarding the history of the current structures and past structures that often cannot be obtained elsewhere. Records of particular interest to the environmental professional are the following:

- Installation, abandonment, and removal of USTs
- Blueprints for building construction, which may include the location of hazardous waste and hazardous material storage areas, facility process areas, heating systems (including USTs and associated piping), and wastewater systems (including septic systems)
- Inspection records of facilities, especially records of violations of building code due to issues related to hazardous material or petroleum storage or use
- Removal of lead-based paint or asbestos-containing materials (if these issues are included in the Phase I ESA)

Engineering and planning—The engineering or planning department often maintains a library of historical aerial photographs (see Section 5.3.4.1). It can also be a source of historical information on public works projects (roadways, water, sewer, electric) or the history on large properties that have since been subdivided.

Fire department—Many jurisdictions require annual inspections by fire inspectors who look for compliance with local fire codes. These people are often valuable sources of information for the Environmental

Professional, not just regarding current practices of storage and usage of organic chemicals and petroleum products, but past practices as well. They are also good sources of information regarding chemical spill incidents, leaking storage tanks, and the like. In many jurisdictions, bulk storage tanks, both USTs and ASTs, must be registered with the local fire department, making the department an important database resource as well.

Health department—Local health departments are generally focused on public health issues, such as vaccinations and disease prevention. However, they may also maintain records regarding USTs, hazardous waste spills, and the like, and they should be contacted as part of the Phase I ESA.

Water and sewer department—The local water and sewer department(s) may have information about when a property was connected to the sanitary sewer system, or when a sanitary sewer line became available to the property. If there is a gap between the time the property was developed and the time when a sanitary sewer line became available, it is likely that a septic tank was once present on the property. Similarly, if the property was not connected to the municipal water system at the time of development, then a drinking water well may have been present on the property. Although its presence does not have the same implications as a septic system, the presence of a drinking water well may present a hazard to building occupants if the groundwater is contaminated, and may provide a convenient path for contaminants to work their way down through the subsurface if present, or, worse, if introduced directly into the drinking water well.

Environmental commission—Various jurisdictions may have environmental commissions to adjudicate local environmental issues that may crop up from time to time. These issues typically involve non-ASTM issues such as wetlands, open space, and traffic. However, these commissions, often composed of volunteers, may have knowledge of property conditions involving hazardous waste or petroleum, especially if the property in question is well known, centrally located, or near a sensitive waterway or other significant environmental receptor.

In many jurisdictions, it is necessary to file a request for records access under the federal Freedom of Information Act (FOIA). It can take up to several weeks to obtain access to these records, so FOIA requests should be submitted as soon as practicable. Even then, this information is often not accessible within the time constraints of the Phase I ESA (see Section 5.4).

5.3.6 Neighboring Properties

Neighboring contaminated properties can impact a particular site via the migration of chemicals. Contaminants on a nearby property could migrate onto the subject property via surface water or overland runoff. The less visible and more common pathway for chemicals to migrate onto a property is via groundwater (see Chapter 4 for a discussion on the principles of groundwater flow).

When groundwater flow data exist and are readily available, it is relatively straightforward to determine whether a chemical spill on a neighboring property has impacted or has the potential to impact the subject property. In most instances, however, that data does not exist or is not readily available to the EP. Therefore, the EP must make certain assumptions in evaluating whether an off-site chemical spill is an REC.

5.3.6.1 Mandatory Database Searches

The E1527 standard identifies seven databases that are managed by the USEPA and list facilities that, at a minimum, are documented to have stored or used hazardous substances or petroleum products. Four state databases are identified in the standard. Table 5.2 lists the databases that must be

TABLE 5.2

Mandatory Public Database Review

Database	ASTM Recommended Search Radius
Federal Databases	
National Priorities List (NPL)	1 mile (1.6 km)
Comprehensive Environmental Response, Compensation, and Liability Information System (CERCLIS)	½ mile (0.8 km)
RCRA Corrective Action Database (CORRACTS)	1 mile (1.6 km)
RCRA Treatment, Storage, and Disposal Facility	½ mile (0.8 km)
RCRA-LQG (Large Quantity Generator)	Site and adjoining properties
RCRIS-SQG (Small Quantity Generator)	Site and adjoining properties
Emergency Response Notification System (ERNS)*	Site only
State Databases	
SHWS (State Hazardous Waste Sites)	1 mile (1.6 km)
SWL/LF (Solid Waste Landfill or Landfills)	½ mile (0.8 km)
Regulated UST (Underground Storage Tanks) Database	Site and adjoining properties
LUST (Leaking Underground Storage Tanks) Database	½ mile (0.8 km)

Source: ASTM.

* The ERNS list is used to collect information on reported releases of oil and hazardous substances. The database contains information from spill reports made to federal authorities including the EPA, the U.S. Coast Guard, the National Response Center, and the Department of Transportation.

researched for an ASTM Phase I ESA and provides the recommended search radius for each database. It is mandatory that properties listed in these databases are assessed for their ability to impact a site.

The issues represented by these databases may range from something as simple as the presence to a UST on the site to something as complex as a site on the National Priorities List (NPL). Generally speaking, the recommended search radius correlates with an increased likelihood that a site on that database has impacted the subject property.

It should not be surprising that NPL sites have the largest recommended search radius. What may be surprising is that the recommended search radius for NPL sites is limited to one mile given the fact, as previously noted, that NPL sites can have a far more widespread impact than one mile. This recommended radius search was settled on by the E1527 standards committee to avoid this research portion of the Phase I ESA from becoming too burdensome. Although there may be cases where NPL sites in which contamination extends beyond one mile could be missed in a radius search, a one-mile search radius would catch most of the NPL sites of relevance. In addition, due to the widespread knowledge of the existence of NPL sites in a given area, it is likely that other means of research, especially interviews with local officials, will enable the Environmental Professional to discover large NPL sites and assess their potential to impact the subject property.

State hazardous waste sites (SHWSs) go under various acronyms in their states, include a cornucopia of different types of spill cases, and therefore constitute varying levels of environmental concern. Because in some states they are the state equivalents of NPL sites, they are given the same recommended search radius as NPL sites. However, it should be noted that this and the other search radii are recommended rather than mandatory. If the Environmental Professional determines that SHWSs in a given state should not be given the same consideration as NPL sites, then a smaller search radius can be used while still conforming to the ASTM Phase I ESA standard.

With the possible exception of SHWS sites, *leaking UST* (LUST) sites are the most prevalent sites on most radius searches. The overwhelming number of leaking UST sites involves petroleum products, either a motor vehicle fuel or a heating oil. A study performed by Lawrence Livermore National Laboratory (LLNL) in 1995 on leaking underground fuel tanks (LUFTS, as they are known in California) concluded that fuel hydrocarbons (FHCs) rarely migrate more than 250 feet from the source. This limited migration distance is due mainly to passive bioremediation occurring at the downgradient end of the plume by microorganisms that recognize benzene and other petroleum products as food to be consumed. From that study, the ASTM E1527 committee considered shortening the search radius for LUST cases.

A development that occurred right about that time put an end to that consideration. The LLNL study focused on benzene, one of the BTEX (benzene, toluene, ethylbenzene, and xylenes) compounds but did not take into account an additive that had just become widespread in the early 1990s, namely, methyl tertiary butyl ether, or MTBE (see Chapter 3). MTBE is not recognized as food by microorganisms, and, due to its chemical composition, has the ability to spread quickly and efficiently throughout an aquifer. With the discovery of LUST sites such as one in Farmingdale, New York, in which an MTBE plume was documented to travel over 1.5 miles from its source, the thinking on LUST was permanently changed. As with NPL sites, LUST sites can impact properties more than ½ mile from their point of origin, but that search radius captures the overwhelming majority of sites that can impact a subject property.

The bane of many a Phase I ESA is the radius search in an urban area, where literally hundreds of sites are present on one or more of the mandatory databases within their recommended search radii. Often it could take several hours to sort through these sites and even more time to evaluate their ability to impact the subject property. The standard allows the Environmental Professional to reduce the search radius for a given database if, in the EP's professional judgment, the deleted information is not significant to the objectives of the Phase I.

Three major databases have a recommended search radius that only entail the site and adjoining properties. Two of these databases, *RCRA large quantity generators* and *RCRA small quantity generators*, deal with the generation of hazardous wastes at the site (see Chapter 2). The generation of hazardous waste implies the storage and usage of hazardous materials or petroleum products at the site. Therefore, the listing of the site or adjoining properties on this database warns the Environmental Professional that it is worth looking into the potential impact of these activities. However, since their presence on one of these databases does not imply the occurrence of a spill of hazardous substances or petroleum, they are considered to have a relatively low potential to impact the subject property "in the absence of a spill report." Future events could change that determination.

The *registered UST database*, usually managed by a state environmental regulatory agency, is the other mandatory database with the recommended search radius limited to the site and adjoining properties. State requirements for UST registration vary but are generally based on content and UST capacity. Registration programs typically do not include the tens of thousands of small heating oil tanks found in single-family residences and small commercial properties. The databases do, however, include numerous USTs, especially in densely populated areas, each having the potential to release its contents directly to the subsurface. USTs on adjoining properties are also of concern because they represent documented petroleum (or possibly chemical) storage near to the subject property; depending on their proximity to the subject property, even a relatively small leak or overfill could impact the subject property.

Since "adjoining properties" vary widely in their distance from a subject property, it is left to the Environmental Professional to decide what search radius would pull in all of the properties that adjoin the subject property. For larger properties or for irregularly shaped properties, it may be necessary to increase the search radius for these databases to ensure that all adjoining properties are evaluated. For small properties, especially properties in densely populated urban areas, it may be advisable, even necessary, to reduce the search radius for these databases as much as possible.

5.3.6.2 Additional Database Searches

In addition to the mandatory databases cited in this section, the federal government and most state governments maintain numerous databases that track a myriad of other environmental activities. Although it is not mandatory to review these databases, they should be perused for information that could be useful to the Phase I ESA. For instance, air emissions data may indicate the usage of a certain chemical, or the generation of a certain waste product that may not have been revealed elsewhere in the document trail. Since the E1527 standard does not specify the recommended search distance for nonmandatory databases, the Environmental Professional must use judgment in selecting an appropriate search radius, based on the nature of the database and its potential importance in identifying RECs on the subject property.

5.4 Limits of Due Diligence Research

The E1527 standard is clear that due diligence research is not meant to be exhaustive. Section 4.5.2 of the standard states:

> There is a point at which the cost of information obtained or the time required to gather it outweighs the usefulness of the information and, in fact, may be a material detriment to the orderly completion of transactions. One of the purposes of this practice is to identify a balance between the competing goals of limiting the costs and time demands inherent in performing an *environmental site assessment* [italics theirs] and the reduction of uncertainty about unknown conditions resulting from additional information.

In general, there are three factors to be considered in deciding whether information, historical or otherwise, is available but would be considered a potential "material detriment to the orderly completion of transactions." The first factor is whether the material is *publicly available*. The E1527 standard does not require the Environmental Professional to obtain information that is not available through public information services. If it is not publicly available, then it is left to the judgment of the Environmental Professional whether the potential usefulness of the information outweighs the time and potential cost of obtaining the information.

A data source is not considered to be *practically reviewable* if extraordinary analysis of irrelevant data would be required to extract the necessary information from the data source. A good example of this second criterion would be telephone books. To find out who occupied a certain building at a certain time, the researcher could go through every page of the phone book, and get the names of companies and people whose numbers were not unlisted. However, this would be an extremely cumbersome method of research, and one that would not be required under the E1527 standard.

Information that is not *reasonably ascertainable*—the third criterion—is publicly available and practically reviewable but not obtainable from its source within reasonable time and cost constraints. If an appointment to review files available at a regulatory agency cannot be made until 6 weeks after the request has been submitted, the information in that file would not be considered reasonably ascertainable. This is often the case with FOIA requests. Similarly, if a bureau intends to charge you several hundred dollars for a review of that file, it would also not be considered "reasonably ascertainable."

Because the E1527 standard is not exhaustive in its scope, it is inevitable that knowledge gaps will remain after the completion of the due diligence activities. These are known as *data gaps*, which the E1527 standard defines as information gaps resulting from an inability to obtain required information despite good faith efforts to gather such information. If the data gap is due to a gap in the site's historical record that is longer than 5 years, it is known as *data failure*.

Here are a few examples of data gaps. The latter two examples are also examples of data gaps:

- Could not access part of the building/property
- Could not interview key person
- Property owner would not release a report on a UST removal that occurred last year
- Could only identify property usage back to 1965
- No historical information between 1950 and 1980

Data gaps by themselves do not doom the Phase I ESA to irrelevance, since some data gaps are much more important than others. A data gap would be deemed "significant" if, based on professional experience, the missing data would have had a reasonable chance of indicating the presence of an REC.

For example, being unable to visit the manufacturing area of a chemical plant would be considered a significant data gap, because the activities that presumably are occurring in that area are likely to have environmental implications. Being denied access to the office area of the same plant would be a lesser concern to the Environmental Professional. Being denied records of a UST removal might be considered significant, but only if that information is not available elsewhere. An interview with the plant manager, who had actual knowledge of the UST removal, may be sufficient to prevent this data gap from being significant.

The significance of historical data gaps has to be considered in the general context of the property history. If the documented history of the property or environs suggest the occurrence of activities that could impact the environment in that time period, then the data gap should be considered significant. For instance, say a property has residential usage, then 25 years later is vacant. Neither of these property uses, by themselves, would be considered a material threat to the environment. If, however, the properties surrounding that property have industrial usage or if, by its location, the property could have contained a filling station, then the data gap could be considered significant.

Even significant data gaps by themselves do not degrade the quality and importance of the Phase I ESA. However, a significant data gap could be as important as an REC and therefore worthy of follow-up action. Sampling and analysis may be conducted to obtain data to address data gaps. In some cases, it may be prudent to do a file review, interview additional people, wait for a response to an FOIA request, or in general dig a little deeper into the documentation before digging into the soil. The E1527 standard does not require follow-up activity; such follow-up is strictly at the discretion of the user (and, in many cases, the user's lender), who must decide whether to live with the uncertainties in the environmental condition of the site. The user, the user's lender, or some other financial stakeholder in the property transaction will usually make this decision.

5.5 Report Preparation

The final product of a Phase I ESA is the Phase I ESA report. This report must include all of the relevant documentation collected in the course of the Phase I activities. The E1527 standard provides a suggested report format; many financial institutions and other frequent users of Phase I ESA reports have other preferred formats. Whatever format is used, the Phase I ESA report must contain, at a minimum, the following elements:

- The identity of the Environmental Professional and the person who conducted the site reconnaissance, if different than the Environmental Professional
- The scope of services performed (it should be noted whether the scope of services included any excluded items, such as an asbestos survey or soil sampling)
- A list of findings, which includes RECs, HRECs, and de minimis conditions
- The Environmental Professional's opinion as to the potential for the conditions listed in the Findings section to impact the subject property
- A list of data gaps and a discussion as to their significance
- The conclusions of the assessment
- Any deviations from the standard practice
- References used in preparing the Phase I ESA report

References

ASTM International. 2005. Standard Practice for Environmental Site Assessments: Phase I Environmental Site Assessment Process (E1527-05).

Hess-Kosa, Kathleen. 2008. *Environmental Site Assessment Phase I: Fundamentals, Guidelines, and Regulations*, 3rd ed. Boca Raton, FL: CRC Press.

Lawrence Livermore National Laboratory. 1995. California Leaking Underground Fuel Tank (LUFT) Historical Case Analysis. UCRL-AR-122207.

U.S. Environmental Protection Agency, Office of Pollution Prevention and Toxics. 1994. Chemicals in the Environment: Perchloroethylene.

6

Site Investigations

As with the Phase I environmental site assessment, the next phase of work goes under several different names. It is often called a Phase II environmental site assessment, especially when performed as a follow-up to the Phase I ESA. In this chapter, we refer to the next investigative phase by another commonly used phrase: the "site investigation."

6.1 Initiating the Site Investigation

There are many reasons to conduct a site investigation. It may be triggered by the discovery of environmental concerns, possibly through a Phase I ESA (although a Phase I ESA does not have to precede a site investigation); a report of a potential release; a real or perceived health impact; the suspected failure of an underground storage tank; or some other incident involving a hazardous chemical or petroleum product.

Financial considerations can also initiate a site investigation. Lenders may want to understand the potential liabilities associated with a prospective loan on a piece of real estate beyond the information provided by a Phase I ESA, especially on industrial properties and other "high-risk" properties, such as filling stations and dry cleaners. Insurance companies may also require site investigations as part of their risk management practices. The passage of the Sarbanes–Oxley Act in 2002 required corporations to disclose financial liabilities in their accounting statements, leading many corporations to attempt to quantify their environmental liabilities.

In some jurisdictions, the site investigation can be triggered by regulations. In New Jersey, for instance, industrial "establishments" that are being purchased or closed must undergo a preliminary assessment, New Jersey's equivalent of a Phase I ESA. Areas of concern identified by the preliminary assessment must undergo a site investigation if there is reason to suspect a release of a hazardous substance or petroleum.

6.2 Developing the Scope of Work

A site investigation can involve one or two areas of concern, all known areas of concern, or no areas of concern in the case of a baseline survey. Regardless of the number of areas of concern and the chemicals involved, many decisions need to be made before data are collected.

6.2.1 Establishing Data Quality Objectives

In a world with unlimited time and money, all investigations could be exhaustive and perfect. Alternatively, in a world where all data are created equal, all investigations could be quick and cheap. In the real world, budget and schedule cannot be ignored, resulting in site investigations ranging in quality between "exhaustive and perfect" and "quick and cheap." We say "between," because even with unlimited time and money, a site investigation will never reach certainty because of the vagaries of the existence of chemicals in the environment and the natural geological variations in the subsurface. Instead, we rely on sampling, which by its very nature leaves gaps in knowledge.

Establishing *data quality objectives* before designing the site investigation guides the investigator on where the environmental study has to fall on the quality spectrum. Since the core of the site investigation is the collection and analysis of samples, data quality objectives will affect the number of samples collected, the effort expended in collecting those samples, the analyses performed on the samples, and the degree of quality control employed in the laboratory analyses.

6.2.2 Conceptual Site Model

The sampling and analysis to be performed will depend upon numerous factors, including, the

- Area of concern(s) (AOCs) to be addressed
- Type of soils or bedrock that are expected to be encountered
- Chemical and physical properties of the compounds that may be present
- Area topography
- Presence of surface waters
- Anticipated groundwater flow direction
- Presence of buildings, subsurface utilities, and other man-made obstructions and preferential pathways in the area of concern

All of these factors can be incorporated into the formulation of a conceptual site model. The conceptual site model can be thought of as the zen of the site. In other words, if I were a chemical detected in a sample collected in an area of concern, how did I get there, where would I go? Samples should be collected at the locations where chemicals are expected to have migrated and at a frequency that provides a level of assurance (not a certainty) that a chemical release, if present, will be detected.

A few examples will help elucidate this point.

EXAMPLE 1: UNDERGROUND STORAGE TANK LEAK

Figure 6.1 is a conceptual diagram of a release from an underground storage tank (UST). Where will the liquids go? That mainly depends upon the chemistry of the liquid and the nature of the geology. If the release occurs in the vadose zone, then in most cases the primary movement of the liquid will be downward. However, there will be some lateral spreading, mainly due to the effects of dispersion, but also due to retardation in the primary direction of migration caused by the presence of organic carbon and other chemical or physical characteristics of the soil. The result of downward movement from the source area combined with lateral spreading results in the liquid spreading out in a cone-shaped pattern in the vadose zone (although the pattern could be asymmetrical if the soils vary laterally). Fine-grained soils will cause more spreading than coarse-grained soils. Similarly, compounds with low carbon–water partition coefficients or low molecular weights will spread out less than compounds with high carbon–water partition coefficients or with high molecular weights, resulting in narrower spreading cones.

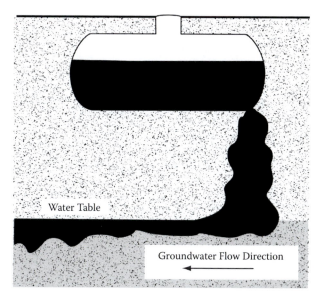

FIGURE 6.1
Conceptual diagram of a petroleum release from an underground storage tank.

Volatile types of petroleum, such as gasoline, will lose much of their mass to the air in the vadose zone; petroleum with low volatility will enter the vadose zone more efficiently. However, the volatile portions that do reach the vadose zone tend to migrate efficiently in the subsurface. Petroleum products generally have low soil–water coefficients, indicating that much of the downward-moving petroleum will be left behind as molecules adhere to soil. This effect creates an area of residual contamination, known as the *smear zone*, that can be detected by soil sampling.

When the liquid reaches the water table, it will dissolve, unless it has a high octanol–water coefficient, as is the case for most types of petroleum. If the contents of the UST were a light, nonaqueous phase liquid (LNAPL), such as no. 2 fuel oil, then the liquid would mainly float on the surface of the water. Whether in nonaqueous phase or in dissolved phase, the liquid would move through advection primarily in the direction of groundwater flow, while spreading out due to the effects of dispersion and diffusion.

EXAMPLE 2: PERCHLOROETHYLENE SURFACE SPILL

A surface spill of perchloroethylene (PCE) will often volatilize before reaching the soil, although it is notorious for finding its way into the subsurface. Once in the subsurface, PCE, like most chemicals, moves

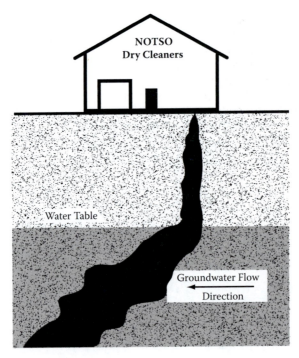

FIGURE 6.2
Conceptual diagram of a petroleum release from a dry cleaning store.

primarily downward through the vadose zone (see Figure 6.2). There will be some conical spreading due to the geology of the soils, but with less conical spreading, due to its very low organic carbon partition and soil–water coefficients. PCE remaining in the vadose zone can result in a vapor intrusion issue if there is an occupied building nearby (see Chapter 7).

When the PCE reaches the water table, it will dissolve to the degree allowed by its solubility, and move via advection, dispersion, and diffusion in the saturated zone. Beyond its solubility point, PCE, which has a specific gravity greater than 1, will be a dense nonaqueous phase liquid (DNAPL), moving downward through the water column unless it reaches a physical barrier, usually an aquitard.

EXAMPLE 3: SPILL OF TRANSFORMER OIL CONTAINING PCBS

The primary chemical class of concern in a spill of oil from a pre-1979 electrical transformer is polychlorinated biphenyls (PCBs) (see Chapter 3). PCBs, while soluble in petroleum, are highly insoluble in water. This factor, combined with their very low soil–water partition coefficient, means that they will not move very far in the subsurface, unless they can be transported as a dissolved constituent in petroleum (see Figure 6.3). With a very high vapor pressure, they are unlikely to go into vapor phase in the vadose zone to an appreciable extent.

If by chance PCBs do reach the water table, only a small amount will be transported by the water due to their very low solubility. With a specific gravity greater than 1, they will act as a DNAPL and move downward through the water column, to the extent that they move at all.

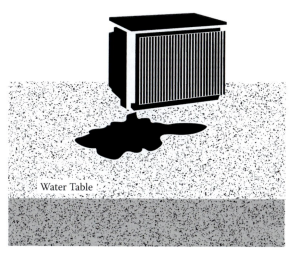

FIGURE 6.3
Conceptual diagram of a PCB release from an electrical transformer.

6.2.3 Sampling and Analysis Plan

The quantity and location of the samples to be collected, and the analyses to be performed, are based on the data quality objectives, the conceptual site model, and the goals of the investigation.

As a rule, the more data available for a given AOC, the more successful the investigation. When less is known, more data are needed. Such sites often merit a broader, less focused investigation than sites where lots of data were available to create a detailed conceptual site model. Table 6.1 lists some common businesses, the contaminants most often associated with those businesses, and the types of analyses that can be used to investigate for those contaminants.

Once decisions have been made regarding the location and number of samples and the analyses to be performed, they can be incorporated into a sampling and analysis plan (SAP). The SAP will provide information regarding the site and site history, including a summary of environmental information collected to date; a description of the project organization (project

TABLE 6.1

Contaminants Associated with Common Businesses

Business	Types of Contaminants	Typical Analyses
Auto service	Petroleum, solvents	PHC, VOC
Auto washing	Petroleum, PAH	PHC, PAH
Auto wrecking/junkyard	Petroleum, metals	PHC, metals
Dry cleaners (on-site cleaning)	perchloroethene	VOC
Foundry	Metals, solvents	Metals, VOC
Gas station (current)	Petroleum, gasoline	PHC, VOC
Gas station (old)	Petroleum, gasoline, lead	PHC, VOC, lead
Machinist	Metals, solvents, petroleum	PHC, VOC, metals
Metal plating	Metals, solvents	Metals, VOC
Painter	Solvents, metals	Metals, VOC
Photo finisher	Solvents, metals	VOC, phenols, silver, zinc
Plastic fabrication	Solvents, metals	VOC, SVOC
Printer	Petroleum, solvents, metals	PHC, VOCs, silver
Railroad	Petroleum, PAH, solvents, paint, fungicides, insecticides	PHC, PAH, VOC, pesticides/PCBs
Sheet metal works	Metals, solvents	PHC, VOC
Welding	Metals, solvents	PHC, VOC

Source: Adapted from Alaskan Way Viaduct Replacement Project: Final Environmental Impact Statement and Section 4(f) Evaluation, Final EIS.

Note: PAH, polycyclic (or polynuclear) aromatic hydrocarbons; PCBs, polychlorinated biphenyls; PHC, petroleum hydrocarbons; SVOC, semivolatile organic compounds; VOC, volatile organic compounds.

TABLE 6.2

1313 Mockingbird Lane, Dogpatch, USA, Sampling and Analysis Plan

Sample Name	Media	Location	Depth Below Grade	Analyses
S-1	Soils	South of UST	10'–12'	TPH-DRO
S-2	Soils	West of UST	10'–12'	TPH-DRO
S-3	Soils	North of UST	10'–12'	TPH-DRO
S-4	Soils	East of UST	10'–12'	TPH-DRO
S-5	Soils	PCE waste storage area	0'–2'	VOCs
S-6	Soils		10'–12'	VOCs
S-7	Soils	Electrical transformer	0'–2'	PCBs, TPH-DRO
MW-1	Groundwater	Near UST	Below water table	TPH-DRO

manager, field supervisor, etc.); the data quality objectives; the conceptual site model; field methods to be employed; and quality control procedures.

The heart of the SAP is a table that provides the sampling locations, the media to be sampled, and the analyses to be performed. An example of such a table is shown in Table 6.2 for a fictitious manufacturing plant (see Figure 6.4).[*] In Table 6.2, three areas of concern are being addressed: a UST, a waste storage area, and an electric transformer. Four soil samples are proposed around the UST, one from each side of the tank. Knowing that the tank contents are no. 2 fuel oil and the soils are comprised of silty sand, the depth of the samples (10 to 12 feet below grade) should be deep enough to detect a leak of fluids from that side of the tank.[†] The analysis to be performed, total petroleum hydrocarbons (TPH-DRO), emphasizes the diesel range organics within the petroleum scan, which is appropriate to detect no. 2 fuel oil. In addition, a groundwater sample will be collected from a permanent monitoring well, MW-1, that will be installed near the UST (see Section 6.6.1 for a description of a permanent monitoring well).

Two soil samples are to be collected in the PCE waste storage area. A sample is to be collected near the surface since the storage is at the surface, and the shallowest soils are the ones most likely affected by a surface spill. However, because PCE travels readily through the soil column, a soil sample will be collected deeper as well. At this AOC the full Target Compound List (TCL) volatile organic compound (VOC) scan can be performed, or a subset of VOCs that includes PCE.

---

[*] Please note that state and local regulations may dictate specific analyses or types of analyses, as well as the number and location of samples to be collected. The following discussion is for illustrative purposes only. Local rules and regulations should be reviewed prior to the design and implementation of a sampling and analysis plan.

[†] The reader should note that environmental reports invariably provide distances and areas in English units, even though other measurements, such as concentrations, are reported in metric units.

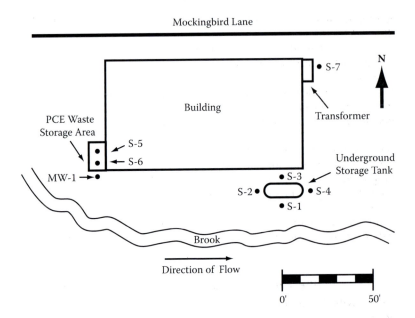

FIGURE 6.4
Schematic diagram of a fictitious factory.

The main concern regarding the electrical transformer is PCBs, so a near-surface soil sample will be collected and analyzed for PCBs. However, the dielectric fluids inside the transformer are petroleum-based, so a TPH-DRO analysis of the soil sample is also warranted.

6.2.4 Sample Analysis

A significant consideration in designing the sampling and analysis program is the type of equipment to be employed for the sample analyses. The data quality objectives, with consideration to time and budgetary constraints, will influence the type of equipment used for the analyses. Some common types of laboratory analyses used in site investigations are described next.

6.2.4.1 Fixed-Base Laboratory Analysis

Protocols for analyses performed by *fixed-base laboratories*, that is, laboratories with permanent locations, are established in the U.S. Environmental Protection Agency (USEPA) publication titled "Test Methods for Evaluating Solid Waste (SW-846)." This document is available at http://www.epa.gov/epawaste/hazard/testmethods/sw846/online/index.htm.

Quality control procedures used in fixed-base laboratory analysis and in laboratory documentation vary widely. The Superfund program requires all laboratories to conform to the Contract Laboratory Program (CLP), which calls for the employment of an elaborate system of quality assurance/quality control (QA/QC) checks. If the data need to comply with federal or state regulatory requirements, then nothing less than a federal- or a state-certified laboratory will suffice. Each regulatory agency dictates the level of quality control required for the laboratory analyses.

Quality control procedures employed by laboratories include the following:

Regular equipment calibrations—Calibrations before and after a series of measurements ensures that the readings provided by the machine are within an acceptable tolerance of accuracy.

Analysis of various blanks—As their name implies, blanks should lack chemicals of interest. If the analytical results for a blank indicate the presence of one or more chemicals of interest, then cross-contamination between samples in the laboratory is suspected. Just as field blanks are important to quality control in the field (as described later in this chapter), so are laboratory blanks important to quality control in the laboratory.

Spike recoveries—A known quantity of a chemical is added to a blank. The sample is analyzed, and the result is compared to the quantity that was added to the blank.

Matrix spikes—A known quantity of a chemical is added to an environmental sample that has been analyzed for a given chemical. The result of the analysis should equal the amount of the chemical in the environmental sample plus the amount of the spike.

6.2.4.2 On-Site Analysis

Mobile laboratories are often used when quick turnaround time is needed, which is known as rapid site characterization. Mobile labs, often located in a field trailer, often have the appropriate regulatory certifications, so that the data can be used to fulfill regulatory requirements. Generally dedicated to one project at a time, sample analysis is usually more expensive than at fixed-base laboratories. However, by enabling the field investigator to get rapid feedback on subsurface conditions, a mobile laboratory can more than make up for the additional expense by saving time and unnecessary drilling and sampling costs.

Laboratory test kits, such as the test kit shown in Figure 14.6 can provide rapid analysis of environmental samples. Within an hour or two, analytical data with a reasonable degree of reliability can be obtained. These types of analyses enable the consultant to make quick decisions in the field. By changing the conceptual site model quickly, days, weeks, and even months can be cut out of the project schedule at little additional cost to the project.

Once the sampling and analysis plan is developed, the consultant should confirm that the proposed plan is consistent with the conceptual site model. If site conditions, anticipated geological conditions, or other variables suggest that the proposed sampling is not sufficient to evaluate the areas of concern, then the plan needs to be revisited and modified, or more research is warranted to obtain the needed information to fill the data gaps.

6.3 Preparing for the Site Investigation

6.3.1 Health and Safety Considerations

The U.S. Occupational Safety and Health Administration (OSHA) regulates the protection of workers at hazardous waste sites under 29 CFR 1910.120. This section of the OSHA regulations is known as the Hazardous Waste Operations Emergency Response Standard, more commonly called by the acronym *HAZWOPR* (pronounced haz'-whopper).

While at the beginning of a site investigation it may not yet have been established that the property is contaminated, it should be assumed that contamination is present for the purpose of worker safety. Therefore, workers conducting a site investigation should have HAZWOPR training. HAZWOPR is a 40-hour training course whose primary intent is to protect workers from uncontrolled exposures to hazardous wastes. Workers receive training in the recognition of hazardous materials, understanding the health effects from exposure to these materials, and understanding the precautions to be taken to prevent exposure.

HAZWOPR requires the preparation of a site-specific health and safety plan, or HASP, at each suspected hazardous waste site. The HASP contains information, such as:

- Potential chemical, physical, and biological hazards at the site
- Lines of communication at the site
- Protective measures to be taken to protect worker health and safety
- What personal protective equipment, or PPE, each worker must use to guard against chemical hazards

PPE generally falls into two categories: dermal protection and respiratory protection. To protect their skin from exposure to hazardous wastes, workers can wear protective suits, usually made of Tyvek™, an inert plastic that has numerous other commercial applications. Workers can also wear protective gloves, sometimes more than one type of glove on each hand for extra protection from chemical exposure. The protective suit is usually taped to the gloves and the worker's boots to complete the skin protection. When more

FIGURE 6.5
A worker in Level B personal protective equipment sampling the contents of a chemical drum.
(Courtesy of the U.S. Environmental Protection Agency.)

dermal protection is needed, workers will wear fully encapsulated suits, affectionately known as "moon suits" due to their similar appearance. With no tape to accidently peel off and expose worker's skin, these suits provide the highest level of dermal protection to the hazardous waste worker.

Respiratory protection comes in two basic varieties. An air-purifying respirator (APR) filters outside air before it is inhaled. In most APRs, the air is filtered by one or two cartridges that are designed to filter out a variety of inhalation hazards, the two most common being organic vapors and particulates. A supplied-air respirator (SAR) bypasses the problem of inhalation hazards by supplying certified clean air to the worker. The clean air is stored in compressed gas cylinders, which are either carried on the back of the worker or located elsewhere and connected to the respirator via flexible hosing (see Figure 6.5).

When the chemical hazards on a site are unknown, OSHA requires workers to wear what is known as Level B PPE, which includes all skin with protective clothing and an SAR. Level C entails dermal protection and an APR if inhalation hazards are expected to be present but not severe. Level D, which involves some dermal protection and no respiratory protection, is typically worn at hazardous waste sites where the chemicals that may be encountered are relatively benign or at sufficiently low concentrations to not present an inhalation hazard to the worker. Such sites make up the overwhelming majority of hazardous waste sites in the United States. Level A protection, used at especially hazardous sites, entails an SAR for respiratory protection and a "fully" encapsulating suit that lacks seams or taped edges that can come apart during use.

Since it is impossible to anticipate the unexpected, OSHA requires air monitoring of site conditions for workers wearing less than Level B PPE. This is critical since the inhalation pathway is typically the greatest threat to the safety of the hazardous waste worker.

Breathing air is generally monitored for organic vapors or particulates. Two types of instruments are generally used to test the levels of organic vapors in the breathable air at hazardous waste sites. The most commonly used organic vapor monitor is the *photoionization detector*, or PID, which provides one number for the total concentration of volatile detected in the air. The *flame ionization detector*, or FID, is more difficult to use but can provide chemical-specific information to the user.

Dust meters measure the amount of particulates in the breathable air. Of primary concern with airborne particulates is the presence of hazardous chemicals attached to the dust, making it an inhalation hazard. Other air meters, such as carbon monoxide meters, combustible gas indicators, and the like are designed to address specific site hazards. Chapter 17 discusses these types of air monitoring equipment.

6.3.2 Utility Markouts

The danger posed by buried utilities, especially electric lines and natural gas lines, is a serious hazard in subsurface work. In the United States, contractors that will penetrate 2 feet or more into the subsurface are required to perform what is known as a utility clearance. In a utility clearance, the various public utilities that have bury utility equipment in the given locale are notified of the impending subsurface investigation; the utilities then send specially trained personnel to locate and mark out the buried utility lines (see Figure 6.6). Different colors signify different buried equipment. There are limitations to the degree that these markouts can be relied upon, but woe to the contractor who does not heed the brightly colored markings.

Unfortunately, public utility personnel will only perform markouts on public property. In some cases, it may be known that, for example, one or more underground storage tanks are present, but their exact location is ambiguous. To obtain more information about buried objects and lines, often a geophysical survey is performed.

6.3.3 Geophysical Surveys

Geophysical surveys involve using remote sensing devices to detect and discern the nature of geological and man-made features without having to

FIGURE 6.6
Utility markouts in a street. The markings indicate the presence of a buried 1¼-inch natural gas line and a buried ¾-inch water line.

actually encounter them directly. There are many advantages to geophysical surveys:

1. *They are safer than direct contact methods.* Geophysical tools can detect buried utilities on private or public property without risking a chance encounter with electric lines, gas lines, and so forth.

2. *They can find buried objects and other items of environmental interest.* Geophysical surveys are commonly used to locate USTs and other buried metallic objects, and former locations of USTs.

3. *They can provide geological information.* This information can then be used to design your site investigation.

Geophysical surveys can be conducted either from the ground surface or at depth in boreholes. Ground surface geophysical surveys are discussed in this chapter; borehole geophysical surveys are discussed in Chapter 7.

The two most commonly used geophysical tools are the magnetometer and ground penetrating radar. *Magnetometers* are most commonly used to detect buried metal objects. They use *electromagnetic* (EM) methods, in which an electrical current generated by the geophysical equipment induces a current in the buried metallic object. The signal generated by the metallic object is detected by the magnetometer. Magnetometers can be set to different sensitivities, depending upon whether the object of interest is large or small,

FIGURE 6.7
Performing a GPR survey on a sidewalk ouside of an office building.

or buried shallow or deep. Magnetometers are also adept at detecting large changes in geology and the depth to the water table, since water is a good conductor of electricity.

Ground penetrating radar (GPR) uses radio waves to detect density variations in the subsurface (see Figure 6.7). Radio waves sent into the surface will reflect off the interfaces of two media with different densities. The greatest density variations are present at solid–gas interfaces. In the subsurface, free gas can only exist within a solid structure. Unless karst topography is present, that structure will usually be man-made, for example, an empty UST, an empty septic tank, or some other empty container. When these structures are filled with liquids, the GPR signal will reflect off their interface with the overlying soils, albeit more weakly due to the smaller density contrast between a solid and a liquid. Unlike the magnetometer, GPR can detect USTs and other buried objects that are nonmetallic. As with the magnetometer, the instrument can be adjusted to detect large, deep objects or small, shallow objects.

Electrical resistivity works on many of the same principles as EM. An electric current is sent into the earth, and signals are sent back to the machine. The strength of the signals depends upon the resistivity of the formation. This geophysical method is especially useful in locating low resistivity geologic formations, such as those that contain water (water-bearing fractures, karst topography, and other preferred groundwater pathways), and high-resistivity formations that do not contain water, such as clay-rich layers. Electrical resistivity can sometimes detect a sufficiently thick layer of nonaqueous liquid, such as petroleum, which has a high electrical resistivity.

Seismic methods make use of acoustic (sound) waves that are sent into the formation. This geophysical method is used to determine the depth, thickness, and composition of geologic strata, the depth to groundwater, and the location and orientation of fractures.

There are two types of seismic methods: refraction and reflection. *Seismic refraction* obtains information from acoustic waves that travel between two geologic layers with differing densities. *Seismic reflection* obtains information from acoustic waves that reflect off these interfaces. These methods are used primarily to obtain information from deeper strata; they are rarely used for investigations of stratigraphy that is less than 50 feet in depth. Acoustic waves are generated by a percussive source, typically a sledge hammer or an explosive charge. An array of listening devices is set up in the area of interest to detect the refractions or reflections from the underlying geologic formations. These listening devices, known as geophones, record the acoustic data, which are then uploaded into a computer and interpreted using specially designed software.

6.4 Soil Sampling

There are three basic investigation methods for soils: test pit installation, borehole installation, and soil gas survey.

6.4.1 Test Pit Installation

The most direct method of soil exploration is the installation of test pits using excavating equipment. An excavator contains a large bucket at the end of an articulated arm (see Figure 6.8). It can remove large quantities of soil in a short period of time. Its smaller cousin, the backhoe, contains a smaller bucket at the end of its articulated arm, and often a shovel at the other end to push soil and grade surfaces. Test pits have the advantage of offering a large, clear view of the subsurface in a short period of time. There are, however, disadvantages in test pit installation: it is difficult to determine the depth of the items observed in the test pit; it causes great disturbance to the surface, which is a problem on paved or occupied properties; and there is greater chance of encountering an underground hazard.

6.4.2 Borehole Installation

The generally preferred method of subsurface investigation is via the installation of *boreholes* (also known as *borings*). Shallow soil samples can be

FIGURE 6.8
Installation of test pits using a backhoe, which has an excavating arm and a shovel to move earth and other heavy objects. (Courtesy of GZA GeoEnvironmental, Inc.)

retrieved with a simple garden scoop or a hand shovel. Also commonly used are hand augers (see Figure 6.9), with which the investigator turns the handle, thereby rotating the cutting bit at the bottom of the auger. The cutting bit digs into the soil, and the soil fills the hollow chamber of the cutting bit. This tool is effective to several feet depth in favorable soil conditions.

FIGURE 6.9
Borehole installation using a hand auger.

For boreholes that must penetrate more than a few feet of soil, a mechanical drilling rig is used. Mechanical drilling rigs come in two varieties: the rotary drilling rig and the direct push drilling rig. The rotary drilling rig utilizes technology originally developed for the oil industry. Mechanical drilling rigs can be mounted on the back of a truck; or on tracks like an army tank for areas where the ground is soft, uneven, or marshy, or as stand-alone equipment. A *hollow stem auger rig* utilizes a helical screw, like a hand auger (see Figure 6.10). The hollow stem auger also has auger "flights," which revolve in a counterclockwise manner, bringing the penetrated soils, known as soil "cuttings" or "spoils," to the surface.

In the course of drilling with a hollow stem auger rig, soil samples can be retrieved using a *split spoon sampler* (see Figure 6.11). The split spoon sampler consists of two hollow stainless steel tubes split longitudinally and equipped with a drive shoe and a drive head. They are generally 2 feet or 4 feet in length. The drive head is attached to the drill rig and lowered to the bottom of the existing hole. The drill rig is equipped with a hammer that drives the split spoons into the ground. As they penetrate the soil formation, the hollow barrel fills with soil. Once retrieved, the spoons are separated, exposing the soil sample.

Drillers will typically count the number of hammer hits it takes to drive the split spoon through a 6-inch interval. The number of hits it takes to drive the length of the spoon is indicative of the competence of the underlying soils. Recording *blow counts*, in what is known as a *standard penetration test*,

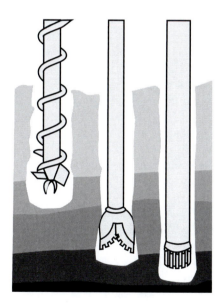

FIGURE 6.10
Different types of drilling equipment. Left: Helical flights of a hollow stem auger. Center: Rotary bit for a rotary drilling rig. Right: Solid stem for a cable tool drilling rig.

FIGURE 6.11
Soils retrieved from a split spoon sampler. (Courtesy of GZA GeoEnvironmental, Inc.)

is used in soil engineering and foundation design to determine the load-bearing capacity of a soil formation.

In this chapter's fictitious factory example, a hollow stem auger rig could drill down to 8 feet below grade at the locations of boreholes S-1 through S-4. At that depth, the driller would disconnect the hollow stem augers and attach a 4-foot-long split spoon sampler to the drive shaft. The rig's hammer would drive the shaft into the ground until it has reached a depth of 12 feet. For boreholes S-5 and S-7, the driller would not use the hollow stem augers at all but rather attach a 2-foot-long split spoon at the surface and hammer down 2 feet into the ground.

In bedrock or hard soil formations, more horsepower is needed than can be provided by a hollow stem auger rig. For such formations, an *air rotary drilling rig* is used (see Figure 6.10). The air rotary rig sends compressed air through the hollow stem augers, which helps lift the cuttings onto the flights so that they can be brought more easily to the surface. By doing so, the auger bit can concentrate more of its energy on cutting through rock and hard soil rather than grinding on material that it had already cut. One drawback of the air rotary drilling method is that its pulverizing and blowing actions make it difficult to determine the type of rock or soil being penetrated, or whether the water table has been reached.

For even more competent and deeper formations, a *mud rotary rig* often is used. The mud rotary rig sends specially designed mud through the hollow stem augers, which helps to cool the drill bit, so that more power can be supplied to the drill stem without the risk of overheating and failing. This technology, widely used in the oil industry, has limited applications in

FIGURE 6.12
Borehole installations using a GeoProbe™. (Courtesy of GZA GeoEnvironmental, Inc.)

environmental drilling, since the mud makes it difficult to obtain geologic information or information regarding the location of the water table.

The second type of drilling technology is the percussion method, in which a tool pounds the ground, gutting out a hole. A cable-tool drill rig (shown in Figure 6.10) operates by lifting and dropping a heavy steel drill "string" into the ground. This drilling method can install holes of over 1000 feet without the use of mud, making it easier to detect groundwater.

A percussion drilling method developed especially for the environmental industry involves using a *direct push drilling rig* equipped with a solid rather than a hollow stem (see Figure 6.12). For this drilling technology, a solid stem of steel is hammered or vibrated down into the ground. Generally referred to as a GeoProbe™, after the firm that invented the technology, the direct push drill rig has the advantage of creating a much smaller hole than a hollow stem auger, which is desirable for developed properties. It is also smaller and lighter than a hollow stem auger rig, so it can manuever into tighter places than a hollow stem auger rig (the smallest rig manufactured by GeoProbe is 62 inches [1.32 meters] tall and 23 inches [0.58 meters] wide). The GeoProbe rig uses clear acetate liners, which functions in a similar manner to split spoon samplers in retrieving soil samples from a given depth interval.

Site conditions often dictate the type of investigation equipment that is used. Some equipment cannot work in areas with limited access; have limited depth range; or cannot penetrate hard rock, concrete, or other known obstructions. Natural obstructions include water bodies, steep slopes, wetlands, and soft ground. Man-made obstructions include low ceilings inside buildings, narrow alleys, and the like. Heavy equipment also must avoid getting near overhead electrical lines, which could result in electrocuting the equipment operator.

6.4.3 Soil Observations and Sampling

The first view of the soil samples is a critical time in the site investigation, as the field geologist looks for indicators of contamination. Is there a chemical odor emanating from the soils? Do they appear stained or otherwise appear unnatural? After the initial inspection, the field geologist should break up the soil sample at various intervals and analyze each interval separately. Each interval will be assessed for the presence of indicators of contamination using visual and olfactory evidence, and interval readings from an organic vapor meter. The field geologist also makes note of the soil types encountered and describes them using the classification techniques discussed in Chapter 4.

Soil samples that will be sent to a laboratory for analysis are collected using decontaminated sampling equipment. Field indicators of contamination often provide new data to the investigator, possibly forcing a reevaluation of the conceptual site model and requiring the collection and analysis of additional soil samples, thus modifying the sampling and analysis plan. Once samples are collected and the desired information is obtained from the borehole, it is backfilled and abandoned, unless it is needed for another purpose, as discussed later in the chapter.

Contamination in soils is inherently heterogeneous, varying by soil type, depth, and so on. Since the sample that is sent to the laboratory for analysis should be representative of the interval sampled, soils sometimes are mixed in a decontaminated bowl so that a homogenized, presumably representative sample can be sent to the laboratory for analysis. Homogenization is never performed for VOC analyses, since VOCs will volatilize and be lost during the mixing process.

Some regulatory agencies allow soil samples to be *composited*, which involves combining *discrete samples*, that is, samples collected from a specific areal location and depth, into one sample for laboratory analysis. Sample compositing saves money on analytical costs but should only be performed if the discrete samples that comprise the composite sample are expected to have similar chemical composition. Compositing soils with differing degrees of contamination can cause the investigator to miss important clues regarding site contamination.

6.4.4 Recording Field Observations

Field observations of a soil boring are recorded on a *boring log*. The header of the boring log shown in Figure 6.13 contains information about the date, site

| ABC CONSULTANTS | | | 1313 Mockingbird Lane | | BORING NO. | S-1 |
| PROJECT: TOOL & DIE MFG. INC. | | | Dogpatch, USA | | SHEET | 1 of 1 |

DRILLING CO.	Munster Drilling	BORING LOCATION	South of UST		
FOREMAN	H. Munster	GROUND ELEV.	242 ft.	DATUM	None
GEOLOGIST	T. Howell, III	DATE START	4/14/12	END	4/14/12

Drilling method: Hollow-stem auger, split-spoon sampling

| Depth (ft) | SAMPLE | | | Sample Description | Stratum Description | PID Reading | Field Testing |
	Split Spoon #	Penetration/ Recovery	Blows/ 6"				
		24/24	13				
	1		11			0.0	
		24/24	12				
4			10	Light tan, sandy silt	SILT	0.0	
		24/0	8				
	2		10			0.0	
		24/12	15				
8			6			0.0	
		24/24	9				
	3		6	Brown medium-grained sand	SAND	0.0	S-1
		24/24	7				
12			5			0.0	

FIGURE 6.13
Example of a boring log.

name, driller, borehole location, observing geologist, and drilling method. The subsurface strata are described in detail, as are any field indicators of contamination. For each 2-foot sampling interval, the log notes the amount of soil recovery. For instance, for boring S-1, 12 inches of soil were recovered in the 6-to-8-foot interval. In the next column, blow counts are recorded in 6-inch intervals. The soils are described in the next column, and a one-word description of the most prominent soil type is provided in the next column. The two columns to the right of the log record PID readings for the sampled intervals.

6.4.5 Field Quality Control

Since so much depends upon the quality of the data collected in the field, the USEPA and most states require some level of field quality control procedures.

6.4.5.1 Decontamination

Decontamination of field sampling tools prevents contaminants from one sampling location from adhering to the sampling tool and contaminating the next sample collected with that sampling tool. *Cross-contamination* can result in increased investigation and remediation costs, since it may cause the investigator to misinterpret a clean sampling location as a contaminated sampling location.

When decontamination is performed in the field, it generally follows procedures proscribed by the USEPA or another appropriate regulatory agency. It usually consists of at least three steps of cleaning to remove gross contamination (dirt and other visible debris), inorganic contaminants, and organic contaminants. The field sampling equipment cleaning and decontamination procedures outlined by ASTM and recommended by the New Jersey Department of Environmental Protection are as follows:

Part 1: Remove Dirt and Other Visible Debris
- Laboratory grade glassware detergent plus tap water wash
- Generous tap water rinse
- Distilled and deionized water rinse

Part 2: Remove Inorganic Contaminants
- 10% nitric acid rinse
- Distilled and deionized water rinse

Part 3: Remove Organic Contaminants
- Acetone rinse
- Total air dry or pure nitrogen blow out
- Distilled and deionized water rinse

6.4.5.2 Field Quality Control Samples

To test the quality of the field data, a series of quality control procedures are typically performed. One type of quality control sample, known as the *trip blank*, consists of laboratory-prepared, chemical-free water stored in an inert glass jar. It is designed to measure the possible cross-contamination of samples during shipping to and from the site. The analysis is typically for volatile organics and only when environmental samples are of an aqueous matrix. This bottle travels with the environmental samples and remains sealed until it arrives at the laboratory.

If the trip blank is contaminated, the contamination could only have occurred at the laboratory or within the cooler, where a contaminated sample released volatile vapors that were then absorbed by the water in the trip blank. Either way, a contaminated trip blank implies that the field samples it accompanied could also have been cross-contaminated. If the contaminant present in the trip blank is also present in one or more of the field samples, then its origins are suspect, and the data quality is considered suspect as well. Such contamination is usually marked as "qualified," and this data degradation is taken into account when the laboratory data are interpreted.

Another type of blank sample, known as a *field blank*, is designed to test for contamination emanating from the sampling equipment. Laboratory-prepared, chemical-free water is poured over one of the sampling tools and into an inert glass jar. If there are contaminants on the sampling tool, they will be rinsed by the water into the glass jar, where they will be detected by the laboratory analysis. If field decontamination is performed, then the field blank should be performed over a piece of field-decontaminated equipment. It is not unusual for the field blank to contain nitric acid, acetone, or some other solvent used in the decontamination process. The presence of contaminants in the field blank will also result in a diminution of the data quality and affect the interpretation of the laboratory data.

Another method of field quality control is the *blind duplicate*. In the field, sampled material is placed into two different jars. The duplicate sample is a single "blind" sample in that the laboratory receiving the sample would not know which field sample that it duplicates. However, the field geologist knows. Assuming the sample is a true duplicate of the original, that is, the two samples are homogeneous, the analytical results should be identical within the tolerances of statistical variation. If the analytical results for the blind duplicate vary significantly from the duplicated field sample and sample heterogeneity can be ruled out as the cause, it calls into doubt the accuracy of the laboratory analyses.

6.4.6 Sample Handling Procedures

When environmental samples are sent to an analytical laboratory, they are accompanied by a completed *chain of custody*. The chain of custody provides basic information about the site and the sampling personnel, the types of samples collected, the sample designations, when they were collected, and the analyses to be performed on the samples. It gets its name, however, from the signatures at the bottom of the form. As the samples are handed from the sampler to the laboratory, or to any intermediary, such as the transporter, it is signed by both the person in possession of the samples and the sample recipient. At any time, the last person to sign is deemed to have legal custody of the samples, and the series of signatures forms a chain, which documents that the samples were never left alone where they could be tampered with or otherwise cross-contaminated.

6.5 Soil Gas Investigations

Soil gas surveys involve the collection and analysis of gas from the vadose zone. Since the composition of the gas in the vadose zone is influenced by

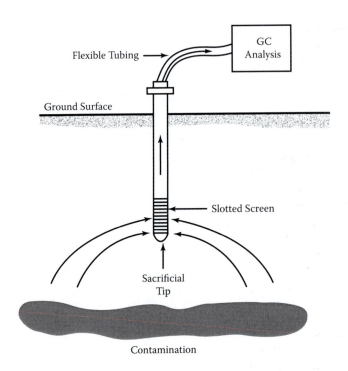

FIGURE 6.14
A schematic diagram of a soil gas probe used in an active soil gas survey.

chemicals in the soils and possibly in the underlying saturated zone, soil gas surveys can be used to assess the presence, composition, source, and distribution of contaminants in the subsurface.

There are two basic types of soil gas surveys. In an *active soil gas survey,* a hollow probe equipped with a screen at the bottom is inserted several feet into the ground (see Figure 6.14). A pump, attached to the probe by plastic tubing, applies a vacuum to the bottom of the probe. The change in the partial pressure in the vadose zone results in volatile contaminants going into the gaseous phase. This gas enters the borehole through a slotted screen and then is pumped up to the surface and collected in a container, usually a flexible container known as a Tedlar® bag. The tip used to drive the probe into the ground remains in the ground (is "sacrificed") while the rest of the probe materials are retrieved from the ground and discarded.

Depending on access issues and site geology, one drill rig can install as many as 30 soil gas probes in a day, enabling the rapid collection of subsurface data. By varying the depth of the soil gas probes, active soil gas surveys can also provide vertical profiling of contaminant concentrations.

Because of their reliance on contaminants readily going into the gaseous phase, active soil gas surveys are most useful at detecting VOCs, especially VOCs that have high vapor pressures and high Henry's law constants.

Active soil gas surveys are only effective in permeable soils, where gas can readily migrate to the gas probes. Clay lenses or clay formations will adversely impact the usefulness of the active soil gas survey. Soil gas surveys also cannot be performed in saturated or near-saturated soils, since soil moisture decreases permeability by blocking vapor flow through the pore space of the soils.

In a *passive soil gas survey*, a sorbent material is placed in the ground and is left there to absorb vapors that are migrating to the surface. Up to several weeks may pass while the sorbent material collects contaminants. This investigation method is most useful for investigations involving semivolatile organic compounds (SVOCs) or when soils prevent sufficient air flow for active sampling. After the probe is removed from the ground, it is sent to a laboratory where the contaminants are desorbed and analyzed.

Passive soil gas sampling is less sensitive than active soil gas sampling to soil permeability or moisture content, although passive gas survey data is also more reliable in high permeability, unsaturated soil conditions. Installing multiple devices at varying depths can provide a vertical profile of the contamination in the area. However, because of the time involved in the field data collection, passive soil gas surveying cannot be used for rapid site characterization. In addition, the data can be adversely affected by time-related variations unrelated to contaminant distribution.

Both active and passive soil gas surveys can be affected by the presence of preferential pathways in the subsurface. Such pathways, including utility trenches, highly permeable backfill, or soil cracks, can cause soil gas to move in an undesirable direction.

6.6 Groundwater Investigations

As with soil sampling, decisions on the method of collecting groundwater samples balance schedule, budget, and data quality, as described next.

6.6.1 Permanent Monitoring Wells

The oldest and most common method for collecting a groundwater sample is by the installation of a *permanent monitoring well*. A permanent monitoring well is installed in a borehole that extends below the water table and into the saturated zone. As shown on Figure 6.15, the bottom of the well contains a *well screen*, usually composed of steel or polyvinyl chloride (PVC). This well screen is set across the water table, enabling formation water to enter the wellbore. Because groundwater elevations tend to fluctuate by at least a few feet over time (and possibly due to tidal influences), the screen is usually situated so that the water table is unlikely to rise above the top of the screen.

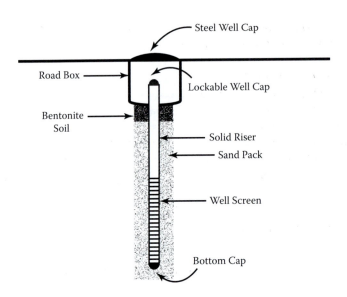

FIGURE 6.15
Diagram of a permanent monitoring well.

This will prevent floating LNAPL from rising above the screen during high stands of water, where it can be overlooked.

The rest of the monitoring well is constructed so that the groundwater sample collected from that well contains a minimum of solids and comes only from the screened interval and nowhere else in the wellbore. A solid plug is attached to the bottom of the screen to prevent infiltration of solids at the bottom of the borehole and a solid riser is attached to the screen to prevent any unwanted solids or liquids above the screened zone from infiltrating into the screened zone. Surrounding the riser/screen assembly is a sand pack that is designed to capture soil grains and other suspended particles. The sand pack is topped with an impermeable material, usually concrete or *bentonite*, a specially designed clay that swells when it comes in contact with water. This solid seal prevents surface infiltration of water and other liquids down the borehole to the screened interval. The top of the well is secured with a cap to prevent unwanted surface infiltration and discourage tampering with the well. When the top of the well is at ground surface it is known as a flush mount. When the top of the well rises above the ground it is commonly known as a "stick-up" mount.

Once the well materials have been installed, water is pumped from the well until the discharge is free of sediment. This process, known as *well development*, removes solids from the wellbore and the surrounding sand pack so that only clear water remains.

Once the well has been developed, the well materials have set into place, and the formation conditions have stabilized, the well can be sampled. Upon opening the well cap, an organic vapor monitor is used to detect vapors in the groundwater. Next, the depth to the top of the water table is measured using a *water level indicator*. Since water levels can fluctuate during well installation, construction, and development, measuring the water level at this time provides the most reliable data regarding depth to water. Water level information is used for a variety of purposes, as is discussed in greater detail in Chapter 7. The water level indicator is usually a flexible ruler with a sensor at the bottom that triggers a sound when water is encountered. A *water/product level indicator* can also measure the top of floating product, typically petroleum, which is useful in determining the thickness of a LNAPL layer in a well.

At this point, a pump is used to remove standing water in the well in addition to solids and other impurities that have accumulated. The standing water is removed because it has had the opportunity to interact with air and is not representative of groundwater conditions at that location. *Conventional well purging* methods call for the removal of three to five well volumes to ensure the complete removal of standing water and the sampling of formation water. *Low-flow purging* methods involve the removal of standing water from a small portion of the screened interval. This purging method has the advantage of generating less purge water that may need off-site disposal if it is contaminated.

During the well purging process, water quality measurements are collected (see Figure 6.16). These measurements have two purposes: to confirm that formation water is being collected; and to obtain information about the groundwater, which can assist in subsequent investigation and remediation activities. These measurements are taken of purge water as it exits the tubing

FIGURE 6.16
Collecting water quality measurements during groundwater sampling. A water level indicator is in front of the field technician. (Courtesy of GZA GeoEnvironmental, Inc.)

rather than in the laboratory, since they can change while the samples are being containerized and transported. They are also used to determine when formation water has entered the borehole, since formation water is expected to have a more consistent and predictable chemistry than standing water in the borehole. These measurements typically include the following:

- Temperature—Formation water that has not been exposed to the air and that is sufficiently deep to be uninfluenced by the air tends to remain at a constant temperature.
- pH
- Dissolved oxygen—Water that has been exposed to air has absorbed oxygen from the air, whereas formation water that has not been exposed to air is often oxygen deficient.
- Turbidity—This measures the amount of suspended particles in the water. A low turbidity sample is desired for reasons discussed earlier.
- Conductivity—This refers to electrical conductivity rather than hydraulic conductivity. Conductivity measures the concentration of ions in the groundwater.
- Oxidation-reduction potential (ORP)—ORP (also known as E_h, measures the ability of a chemical to either donate electrons (oxidation) or acquire electrons (reduction) when dissolved in water. It is measured in volts (V) or millivolts (mV).

Collection of water quality data is particularly critical for low-flow purging, since it provides the only means of determining whether formation water, assumed to be homogeneous, has entered the well. Once the groundwater sample is collected (see Figure 6.17), it is containerized and shipped in a manner described earlier for soil samples.

Another method used to remove suspended particles from a groundwater sample is *field filtering*. Performed in conjunction with either conventional or low-flow purging, field filtering involves pouring the groundwater sample through a microscreen, which captures the suspended particles before they can enter the sampling jar. The disadvantage of field filtering is that chemicals can volatilize due to the extra handling and the impact of the water on the microscreen. Therefore, field filtering should never be performed when sampling for VOCs and is indeed prohibited for VOC sampling in many jurisdictions for that reason.

6.6.2 Temporary Monitoring Wells

Many site investigations utilize *temporary monitoring wells*, also known as temporary well points, rather than permanent monitoring wells. As the name implies, well materials, at a minimum a screen/riser assembly, are

FIGURE 6.17
Collecting a groundwater sample using a plastic bailer. (Courtesy of GZA GeoEnvironmental, Inc.)

placed into the borehole but removed after groundwater samples are collected rather than set into place, as with a soil gas survey. A groundwater sample is collected through the temporary screen, after which the well materials are removed from the borehole.

Temporary monitoring wells are less expensive than permanent monitoring wells and can be installed and sampled more quickly. However, temporary monitoring wells have several disadvantages. They can only be used once, whereas there is no theoretical limit to the amount of times permanent monitoring wells can be used. Also, it is more difficult to keep unwanted sediment and other particulates from entering the temporary wellbore than the permanent wellbore that has been constructed with a sand pack and then developed. This factor is especially important for contaminants that tend to sorb to solids, such as SVOCs and metals. These shortcomings notwithstanding, for rapid site characterizations temporary monitoring wells are indispensable.

6.7 Interpreting and Documenting the Site Investigation

6.7.1 Data Reduction and Interpretation

Table 6.3 is an example of an analytical results table for the soil samples collected at our fictitious manufacturing facility. The table is simplified for illustrative purposes. General information provided in the table includes the sample name, the collection date, and the sample depth. Results are provided for each contaminant class tested, even if there were no detections, such as for TPH-DRO for samples S-1 and S-4. The analytical results are listed and compared to remediation standards, which are discussed in detail in Chapter 7.

In this data set, soil sample S-3 contained TPH-DRO at a concentration above both soil standards. PCE and TCE, which are related compounds, were the only two VOCs detected. Soil sample S-5 contained PCE at a concentration below its remediation standard. However, the deeper sample S-6 contained PCE at a concentration above its remediation standard and some TCE as well. Neither of the soil samples collected around the transformer contained PCBs at a concentration above their remediation standards. If the sampling in this area of concern was adequate, then the investigator will conclude that no further investigation is warranted for this area of concern.

6.7.2 Site Investigation Report

The site investigation report documents the results of the investigation and provides recommendations, if any, for further investigation. The report introduction answers the who, what, where, and whys of the investigation. This introduction includes:

- The objective of the investigation.
- Basic site information (address and sometimes legal description).
- Names of the parties involved, including the site owner and operator, the environmental consultant, and the regulatory agency, if applicable. This section may also describe the project organization, including the names of the project manager, field supervisor, and so forth.
- Background information that guides the reader in understanding the sampling that was conducted. This may include the operational history of the facility, the results of previous environmental

TABLE 6.3

1313 Mockingbird Lane, Dogpatch, USA, Analytical Results Summary for Soils

Sample Name	Remediation Standard	S-1	S-2	S-3	S-4	S-5	S-6	S-7
Date Sampled		7/1/11	7/1/11	7/1/11	7/1/11	7/1/11	7/1/11	7/1/11
Sample Depth		8'–10'	8'–10'	8'–10'	8'–10'	0'–2'	10'–12'	0'–2'
TPH-DRO	5000	ND	50	6200	ND	NS	NS	150
VOCs								
Perchloroethylene	1.0	NS	NS	NS	NS	0.2	4.2	NS
Trichloroethylene	1.0	NS	NS	NS	NS	ND	0.7	NS
PCBs								
Aroclor-1242	2.0	NS	NS	NS	NS	NS	NS	0.49
Aroclor-1248	2.0	NS	NS	NS	NS	NS	NS	0.70

Note: All concentrations are listed in milligrams per kilogram (mg/kg). ND, not detected; NS, not sampled.

investigations and studies, and physical setting information, such as site and area geology, topography, and groundwater depth and anticipated flow direction.

The technical review section is the heart of the site investigation report. This section describes the field activities performed, the field observations, and the results of the sampling and analyses. It includes a map that shows the sampling locations and analytical results tables for each medium sampled, such as Table 6.3. Any limitations in the investigation (obstructions, field equipment failures, etc.) are identified.

Once the results of the field activities, the field sampling, and the laboratory analyses are described, they are summarized in the findings section. For each area of concern, this section identifies the media affected, the extent of contamination (both horizontal and vertical), and the contaminants encountered. The suspected source of discharge is identified and the potential impact of the limitations encountered on the results of the investigation is evaluated.

The conceptual site model should fit all of the data collected for the site investigation. If data does not fit the conceptual site model, then the conceptual site model must be modified (unless there is a sound technical reason why the data may be misleading or incorrect). Evaluation of the conceptual site model continues throughout the remedial investigation (Chapter 7) and the remedial action (Chapter 8) phases of the project.

References

Arkansas Department of Environmental Quality. June 2007. Arkansas Brownfields Program User's Guide.

ASTM International. 1993. Standard guide for soil gas monitoring in the vadose zone, D5314-93, Annual Book of ASTM Standards.

ASTM International. 2002. E1903-97 (2002): Standard Guide for Environmental Site Assessments: Phase II Environmental Site Assessment Process. ASTM International.

ASTM International. 2008. D5088-90 (2008): Practice for Decontamination of Field Equipment Used at Nonradioactive Waste Sites. ASTM International.

Benson, Richard, Glaccum, Robert A., and Noel, Michael R. 1984. Geophysical techniques for sensing buried wastes and waste migration (NTIS PB84-198449). National Ground Water Association.

Driscoll, Fletcher G. 1986. *Groundwater and Wells*, 2nd ed. Johnson Screens.

Kerfoot, Henry B., and Barrows, Larry J. 1987. Soil gas measurements for detection of subsurface organic contaminants, EPA/600/2-87/027 (NTIS PB87-174884).

Lagrega, Michael D., Buckingham, Phillip L., and Evans, Jeffrey E. 2001. *Hazardous Waste Management*, 2nd ed. New York: McGraw-Hill.

New Jersey Department of Environmental Protection. August 2005. Field Sampling Procedures Manual. NJDEP.

Occupational Safety and Health Administration Web site (www.osha.gov).

U.S. Environmental Protection Agency. August 1987. A Compendium of Superfund Field Operations Methods. EPA 540/P-87/001. Office of Emergency and Remedial Response.

U.S. Environmental Protection Agency. March 1997. Expedited Site Assessment Tools for Underground Storage Tank Sites: A Guide for Regulators. USEPA Office of Underground Storage Tanks, OSWER.

7

Remedial Investigations and Setting Remediation Goals

Once contamination has been detected at a site, the next step is to determine the severity and extent of the contamination, with the ultimate goal being its remediation, which is discussed in Chapter 8. The severity and extent of the contamination is evaluated in a *remedial investigation*.

Remedial investigation entails the delineation of contamination in soils, groundwater, sediment, and surface water, as warranted by site conditions. Following is a discussion on the methods used in soil and groundwater delineation (delineation of sediment and surface water is discussed in Chapter 9). This chapter also discusses the identification and delineation of contamination of indoor vapors emanating from subsurface contamination sources.

The remedial investigation step is often combined with the collection of data that will be needed for the design of the remedial action to be performed, including the establishment of remediation standards. This chapter also discusses both the remedial investigation and the steps leading up to the design of the remedial response action to be taken.

7.1 Remedial Investigation of Soils

7.1.1 Delineation of Soil Contamination

Delineation of contaminated soils is usually a relatively straightforward process. The standard practice is to surround the known area of contamination with boreholes, collect soil samples from those boreholes, and to analyze those samples for the contaminants of concern (soil gas surveys can be used in this manner as well, see Chapter 6). Delineation must be accomplished horizontally as well as vertically. Having identified known or suspected migration pathways for contaminants in the conceptual site model, the investigator may choose to bias the sampling in a particular direction. The distance between sampling locations will depend upon the nature of the contamination, the expected distance the contaminants were suspected to have migrated from their point of origin, and the desired degree of certainty in the data set.

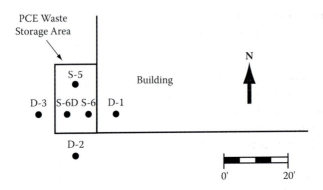

FIGURE 7.1
A map showing the planned delineation of documented soil contamination at a factory site.

Continuing the case example from the previous chapter, the map provided in Figure 7.1 shows the locations of boreholes designed to delineate
the perchloroethylene (PCE) contamination detected in soil sample S-6.
This sample contained PCE at a concentration of 4.2 milligrams per kilogram (mg/kg), with the remediation standard being 1.0 mg/kg. Boreholes
D-1 through D-4 are designed to delineate horizontally the PCE detection
from soil sample S-6. Borehole S-6D, drilled at roughly the same location as borehole S-6, is designed to provide vertical delineation. The soil
samples should be analyzed for a scan that includes PCE, usually one of
the volatile organic compound (VOC) scans.

Field screening techniques can provide useful information when VOCs are
the target of the investigation. For instance, if organic vapor monitor measurements suggest that a soil sample contains high concentrations of organic
vapors, the field geologist may choose not to send that sample to a laboratory for analysis. Since the goal is to identify uncontaminated soils, the field
geologist can assume that this sample is contaminated, and select a new spot
farther away from the original sampling point for sampling and analysis.

Field analytical equipment, such as mobile laboratories and laboratory test
kits, can provide very reliable data in a short period of time, leading to a
rapid characterization of the area of contamination. When such methods are
not used, it is often useful to collect additional soil samples beyond the first
group of samples to be analyzed by the laboratory only if one of the samples
analyzed in the first batch is contaminated.

Alternatively, contamination can be delineated if it can be demonstrated
that a *concentration gradient* existed. A concentration gradient is a consistent
change in concentration over a given distance. If successive soil samples
exhibit a decreasing concentration trend, and this trend is consistent with
the conceptual site model, then the location at which the soils will comply
with remediation standards, known as the *compliance point*, can be estimated. Even with the establishment of a concentration gradient, it still may

be advisable to collect a sample at the calculated compliance point to document the delineation of the contamination. *Compliance averaging* in which the results of samples collected in the same general area are averaged together, is allowed in certain jurisdictions.

7.1.2 Obtaining Quantitative Soil and Bedrock Data

Qualitative data about the soils in the area to be remediated are often insufficient for designing a remedial action for contaminated soils, especially if the soils are to be remediated by means other than excavation. Hard quantitative data are needed.

7.1.2.1 Coring Data from Boreholes

To collect high-quality physical data for soils, it is necessary to obtain a soil sample that has not been disturbed by the drilling and soil collection process. A *Shelby tube* is a hollow chamber that is driven into the soil formation. Once filled, it is capped at the bottom and sent to a laboratory for analysis. The analysis can provide information about the percentage of each soil type and other soil parameters.

As discussed in Chapter 4, groundwater flows mainly through fractures in bedrock. Therefore, understanding the location and orientation of the bedrock fractures is the key first step in understanding the behavior of a contaminant plume in bedrock. In the course of drilling a borehole through bedrock, core samples can be collected. In areas where the bedrock is not fractured, the core will be largely intact. Broken pieces of rock will be present where fractures exist, the degree of fracturing indicated by the quantity of rock fragments of rock present. The percentage of competent bedrock to fractured bedrock is described in the *rock quality designation* (RQD). The RQD provides direct information about the presence and quality of fractures in the rock formation.

7.1.2.2 Borehole Logging

Borehole logging is an indirect, geophysical measuring method that also can provide quantitative soil and bedrock data. It is similar in concept to a surface geophysical surveying (see Chapter 6), except that in borehole logging the measurement devices are lowered into a borehole rather than moved across the ground surface. The measurement device usually is attached by electric wire to a recording device, usually a field computer, located at the surface. In some cases, the tool itself stores the information for eventual uploading onto a computer once it has been retrieved from the borehole. Some of the more common tools used in borehole logging to obtain soil and bedrock information are discussed next.

The oldest and most used logging device is the *electrical log*, which measures the electrical properties of a geological formation. Most electrical logs measure electrical resistivity (*resistivity logs* perform the same function using different equipment). Electric logs can be used to identify saturated versus unsaturated zones, and the type of soils or bedrock present. Dry formations tend to have higher resistivity than saturated formations, and sand and sandstone tend to have higher resistivity than clay or shale.

The *natural-gamma ray log* measures naturally occurring radioactive material, including potassium 40, uranium 235, uranium 238, and thorium 232. Since potassium is a major component of most clay minerals, gamma logs (as they are usually called) are useful in identifying clay formations. The gamma log and the electric log form the common *e-log suite*.

Many of the downhole geophysical tools are geared toward identifying bedrock fractures and their properties. The *caliper log* uses mechanical arms to measure the diameter of the borehole. An increase in borehole diameter within a zone is often indicative of the presence of bedrock fractures or bedding planes. *Downhole imaging logs* include optical televiewers, in which a camera collects real-time images of the borehole, and acoustic televiewers, which produce high-resolution images of the borehole walls by sending out and collecting high-frequency sound waves. Downhole imaging logs can not only identify bedrock fractures or bedding planes, but also provide the three-dimensional orientation, which is critical in the understanding of contaminant transport in complex fractured bedrock formations.

7.2 Remedial Investigation of Groundwater

The process of groundwater delineation is similar to the process for soil delineation, with one important distinction. Groundwater moves, as do the dissolved contaminants in the groundwater (in a similar but not identical pattern). Therefore, understanding this movement and the associated transport of dissolved phase and separate phase chemicals is critical to the delineation of a contaminant plume.

7.2.1 Calculating Groundwater Elevation

Since the movement of dissolved contaminants is similar to groundwater movement, establishing groundwater flow direction is a critical step in delineating groundwater contamination. The fundamental measurement in determining groundwater flow direction is *groundwater elevation*, which is the elevation at which water in the saturated zone is first encountered. Groundwater in a monitoring well or well point (see Chapter 6) is gauged using a water-level indicator, which is a handheld device attached to a spool of measurement tape. When water is encountered, the circuit inside the device is completed, causing the device to emit a sound and a bulb to light (similar devices are available for detecting petroleum and other nonaqueous liquids).

Groundwater gauging points can be obtained from permanent or temporary monitoring wells or through the installation and gauging of a *piezometer*, which is a borehole installed through the water table whose primary purpose is to measure groundwater levels. Monitoring wells can be used both as piezometers and as groundwater sampling points. In the following discussion the term piezometer will apply to all points used for the measurement of groundwater elevation.

Once the depth to water has been measured, the elevation of the water table can be calculated by subtracting this number from the ground elevation of the piezometer, which can be determined by surveying the piezometer. Datum points that tie in to the National Geodetic Datum database are managed by the U.S. Geological Survey (USGS). The datum used today is known as the North American Datum of 1983 (NAD 83), which covers the United States, Canada, Mexico, and Central America. In the United States, datum points are given in feet above mean sea level (MSL). A surveyor will tie the elevation of the piezometer into the nearest datum, thus establishing the elevations of the piezometer and the water table at that location.

The depth to groundwater in a piezometer constantly changes due to seasonal fluctuations, tidal influences, precipitation effects, or for reasons related to microgeologic factors, such as those discussed in Chapter 4. Where there are tidal influences, groundwater elevation and, in some cases, groundwater flow direction will change with the tides. Such fluctuations must be incorporated into the conceptual site model and factored into the remedial investigation.

7.2.2 Calculating Groundwater Flow Direction

The water table derives its name because its top is conceptually like a flat surface like the top of a table. The main difference is that the water table, for purposes of calculation, is a plane that extends an infinite distance in all directions. For groundwater to flow, the table must be tilted in one direction. As geometry teaches us, three points define a plane. Therefore, three groundwater elevation points are needed to define the water table.

Going back to our fictitious contaminated property, Figure 7.2 shows the location of monitoring well MW-1 as well as new monitoring wells MW-2 and MW-3. Well MW-2 was installed in the presumed down-gradient direction and well MW-3 was installed in the presumed cross-gradient direction, thereby forming a triangle with which groundwater flow direction can be calculated.

Groundwater elevations were collected from MW-1, MW-2, and MW-3, and Table 7.1 shows the calculations used to derive their groundwater elevations. The groundwater elevations are then plotted onto the site map to be used to calculate groundwater flow direction. Groundwater flow direction is calculated

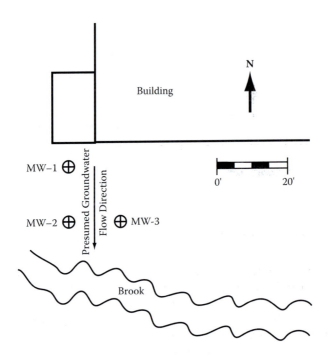

FIGURE 7.2
Groundwater delineation wells MW-2 and MW-3 were installed in the general direction of
the presumed groundwater flow, based on surface topography and the presence of a nearby
stream.

by using triangulation to plot *groundwater contour lines*. Groundwater contour
lines, analogous to contour lines on a topographic map, connect locations of
equal elevation of the top of the water table. By convention, round numbers,
such as the 74.00′ shown on Figure 7.3, define the contour lines.

Since the water table was encountered at 75.00′ in MW-1 and at 73.00′ in
MW-2, there must be a point between these two piezometers at which the
water table elevation is 74.00′ (above MSL is implicit). There are also points
between MW-1 and MW-3 at which the water table elevation is 74.00′ and
73.00′. The location of these points can be found by interpolation. For instance,
the distance of the 74.00′ point from MW-1 along a line connecting MW-1 and
MW-2 is calculated using the following interpolation formula:

$$\left((74.00 - E_1) / (E_2 - E_1)\right) \times D$$

where E_2 is the elevation of the water table at MW-2; E_1 is the elevation of the
water table at MW-1; and D is the ground distance between MW-1 and MW-2.

In Figure 7.3, a similar interpolation process was used to identify the location of the 73.00′ and 74.00′ contour lines between MW-1 and MW-3. These
lines are parallel to each other, as predicted by geometry theory.

TABLE 7.1

Calculation of Groundwater Elevations in Monitoring Wells

Well Name	MW-1	MW-2	MW-3
Top of inner casing	89.00′	86.00′	86.00′
Depth to groundwater	14.00′	13.00′	14.00′
Groundwater elevation	75.00′	73.00′	72.00′

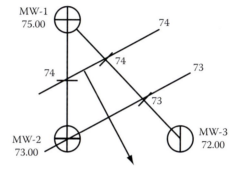

FIGURE 7.3

Contour map of the top of the water table, showing the groundwater flow direction.

Just as a ball will roll off a tilted table in the direction of its maximum tilt, so will groundwater flow in the direction of the maximum inclination of the water table. Therefore, identifying groundwater flow direction is found by drawing a line perpendicular to the groundwater contour lines. The slope of the plane of the water table is known as the *hydraulic gradient*.

A word of caution. The implicit assumption in the aforementioned discussion is that the three monitoring wells are installed in the same aquifer. If one of the three wells is installed in a different aquifer than the other two, an incorrect groundwater flow direction will be calculated. This problem can be avoided by looking out for aquitards that could separate saturated zones, such as happens if a perched water table is encountered in one of the wells. Differences in aquifers can sometimes be detected by looking for significant differences in water quality measurements, such as pH and dissolved oxygen. Sometimes the groundwater contours suggest that different water has been encountered due to an unusually steep calculated hydraulic gradient, for example, or groundwater flow heading in an unexpected direction.

Groundwater flow can also be undermined by the presence of LNAPL floating on top of the water table, which will depress the water table due to its weight.

7.2.3 Contaminant Plume Mapping

Once groundwater monitoring points are installed, groundwater samples are collected and analyzed to determine whether contamination is present at those locations. Table 7.2 provides the TCE concentrations measured in the three monitoring wells from our example. Not surprisingly, monitoring well MW-1, installed near the source of the TCE release, contains the highest concentration of TCE.

These concentrations can be used to develop an *isopleth map* (also known as an *isoconcentration map*), which is a contour map that shows the extent and severity of a contaminant plume in groundwater. The isopleth map contains contour lines that connect points of equal contaminant concentrations rather than equal elevation, as with a topographic map or a groundwater contour map. The contour lines are drawn using interpolation, as described earlier.

The resulting isopleth map, shown in Figure 7.4, shows a contaminant plume that is elongated in the direction of groundwater flow, as is expected when advection is the primary mechanism of contaminant transport. If advection was the only process affecting the size of a groundwater plume, all groundwater plumes would be shaped like a pencil and oriented in the direction of groundwater flow. However, as discussed in Chapter 4, various physical, chemical, and biological processes act to widen the plume. The degree of eccentricity of the ellipse is controlled by the proportional effect of advection versus these countervailing forces.

In summary, the long axis of the ellipse should more or less point in the direction of groundwater flow, and its length is controlled by the hydraulic gradient and hydraulic conductivity; the amount of time since the initial release to groundwater; and the degree to which the leading edge of the plume is degraded by physical, chemical, and biological forces.

It should be noted that, while three wells are required to establish groundwater flow direction, many more wells are often needed due to the effects of subsurface complexities on groundwater flow pattern. This goes doubly true for contaminant plume patterns.

Complex plumes in a complex geologic environment can require numerous wells for a true grasp of the plume and its behavior. Complicated bedrock fracture patterns, such as those shown in Figure 4.8 in Chapter 4, yield complicated groundwater flow and contaminant plume patterns. Complicated fluvial or glacial environments at the time the original soils were deposited, or subsequent alterations of the soils or bedrock, can result in groundwater

TABLE 7.2

TCE Concentration in Monitoring Wells

Well Name	MW-1	MW-2	MW-3
TCE concentration	275 µg/l	5 µg/l	50 µg/l

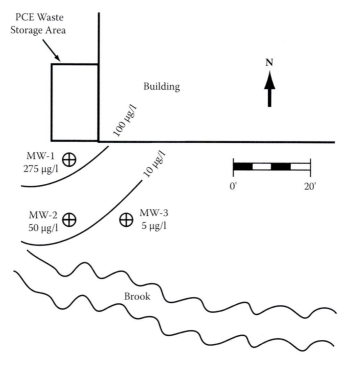

FIGURE 7.4
Isopleth map of PCE in groundwater at a fictitious factory.

contour and isopleth maps that show little if any discernable pattern. Adding complexity to the equation are preferential groundwater flow paths caused by natural and man-made phenomena, and man-made obstacles, such as building foundations. Often, computer modeling, which is described later in the chapter, can assist the geologist in analyzing complex groundwater flow patterns.

As difficult as it is in obtaining an understanding of the two-dimensional structure of a contaminant plume, in reality plumes are three-dimensional. In most cases groundwater contamination must be delineated vertically as well as horizontally, as described next.

7.2.4 Vertical Delineation of Groundwater Contamination

Several physical mechanisms can cause contaminants to migrate downward through a saturated zone. DNAPL can migrate downward because it is denser than the surrounding water. Dissolved chemical constituents, whether denser or lighter than water, will migrate downward within an aquifer primarily through diffusion. Contaminants can migrate into deeper aquifers by means of dispersion via various pathways, including bedrock fractures and limited interconnected points between the aquifers.

Monitoring wells installed for vertical delineation purposes by definition penetrate contaminated zones. It is critical that the contamination from the shallower zone not enter the borehole, because doing so would create an artificial vertical pathway for the contaminants, thus violating the "first, do no harm" principle of environmental investigations. The creation of such an artificial pathway is known as *cross-contamination* and can result in the contamination of previously uncontaminated groundwater. To prevent cross-contamination, the contaminated portions of the aquifer must be sealed off before continuing the drilling activities. This is often accomplished by installing a *double-cased well*, in which a casing is set against the portions of the aquifer above the zone of interest, with a narrower casing set below the upper casing, to which a screen over the zone of interest is attached.

Monitoring wells set at different depths in the same approximate location can enable the geologist to analyze the vertical variations in contaminant concentration. Wells set at different depths in the same physical location are known as a *well cluster*. It is possible to set multiple monitoring wells at different depths within one large casing, which is known as *nested monitoring wells*.

7.2.5 Obtaining Quantitative Groundwater Data: Aquifer Analysis

As discussed earlier, qualitative data about soils to be remediated are often insufficient for designing a remedial action. This is doubly true for groundwater. Quantitative data about groundwater can be obtained from a variety of sources.

7.2.5.1 Aquifer Analysis

Aquifer analysis often is performed as a prelude to the design of a groundwater remediation system. Physical properties of interest in an aquifer analysis include, but are not limited to, the following:

- *Hydraulic conductivity* is the ability of a geologic formation to transmit water. Similar in concept to electric conductivity, hydraulic conductivity is used to estimate the flow rate of fluid through a porous material. It is expressed as a velocity, typically in centimeters per second (cm/sec) in the International System of Units (SI).
- *Transmissivity* is the ability of groundwater to move through an aquifer, given an ideal hydraulic gradient (i.e., vertical). Transmissivity is equal to the hydraulic conductivity multiplied by the aquifer thickness and can be expressed in cubic meters of water per day (m^3/day).
- *Specific yield* is the volume of water that drains from saturated soil as the water table drops. It is a dimensionless number.

- *Storativity* is the volume of water that drains from the saturated soil as the water table drops for an unconfined aquifer. It is dimensionless.

These parameters will enable the geologist to calculate the *potential capture zone*, which is the area in which contaminants may be recovered, or "captured," by a pumping well.

The three most common methods of performing an aquifer analysis are the slug test, the pump test, and the step-drawdown/well recovery test.

A *slug test* is most commonly employed on low-yield aquifers. In a slug test, a volume of water or a solid object (the "slug") is introduced into a well, causing the water elevation to rise in the well. This condition is temporary, as the water level will eventually return to its "static water level" (the level of the water prior to the introduction of the slug), due to the water in the borehole water absorbing back into the geologic formation. The falling of the water level (a "falling head test") is measured either with a water level indicator, or, most often, an electronic data logger. Once the static water level is reached, the test can be concluded. Often, water level measurements are collected after the solid slug is removed and the water level suddenly drops and then rises, in what is known as a "rising head test," to confirm the results of the falling head test. Data obtained is then used to calculate the hydraulic conductivity of the geologic formation.

For high-yield aquifers, the *pump test* is the most common method of aquifer analysis. The pump test is most commonly used for community drinking water wells. In environmental investigations, pump tests are usually conducted as part of the remedial design of a pump-and-treat system, which is discussed in Chapter 8.

As its name implies, in a pump test a water pump is lowered into the well and activated. When water is pumped from a well, forming a *cone of depression* (see Figure 7.5), which is the area that the groundwater vacates during

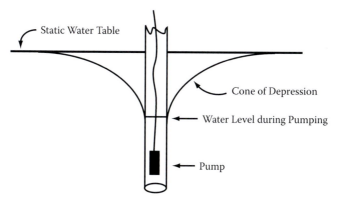

FIGURE 7.5
A schematic diagram of drawdown in a pumping well.

pumping. The change in the height of the water column in the pumping well, known as the *drawdown*, may vary in the course of the pump's operation.

Prior to the pump test, background conditions (static water levels) are established in the pumping well and in nearby monitoring wells. The well is pumped at a constant rate, and the pumping and nearby monitoring wells are continuously gauged. The amount of drawdown that occurs in a given monitoring well over time will be directly related to its distance from the pumping well, provided the geologic formation is *isotropic*, that is, perfectly symmetrical in all directions. A geologic formation that is *anisotropic*, as in the case where an impermeable layer is present only in one direction from the pumping well, will cause unexpected variations in the relationship between a well's drawdown and its distance from the pumping well.

In a *step-drawdown/well recovery test*, a well is pumped at progressively higher pumping rates and the drawdown at each pumping rate is recorded. Usually four to six pumping steps are used, each lasting approximately one hour. The higher the pumping rate, the larger the cone of depression and the greater the portion of the aquifer tested.

7.2.5.2 *Groundwater Data from Boreholes*

Downhole logging tools can collect data regarding the physical properties of the groundwater within fracture zones as well as the physical properties of the rock itself. *Flowmeters* measure water flow velocity between two zones in a non-pumping well. This information can be used in formulating the water flow pattern within a bedrock aquifer. The *temperature log* measures the temperature of the water and can be used to identify water-bearing fractures, establish the geothermal gradient at a location, and correlate between fractures at different well locations.

7.2.5.3 *Computer Modeling*

Computer software can facilitate the understanding and the visualization of contaminant plumes. Understanding the fate and transport of these contaminants can be facilitated by *computer modeling*. Existing geological data can be input into three-dimensional computer models, which can then be used to simulate the flow of groundwater in an aquifer. This information can in turn be used to predict the extent of contaminant plumes, fate and transport of contaminants, potential impact to receptors, and so forth. The de facto standard for computer modeling is a program known as MODFLOW. Free versions of MODFLOW can be downloaded from the U.S. Geological Survey Web site at http://water.usgs.gov/nrp/gwsoftware/modflow.html.

7.3 Geographic Information Systems (GIS)

Data can accumulate rapidly for large remedial investigations. Information about the location of boreholes, the depths of sampling locations, and the results of the laboratory analyses needs to be well-organized to facilitate retrieval and interpretation. Lost data or, even worse, data plotted in the incorrect locations can result in project delays, cost overruns, and even lawsuits, which could happen if the errors result in a failure to remediate existing contamination.

Geographic information systems can assist in the management of site data. At its core, a GIS is a map that is connected to a database or various databases. The map is in digital format, usually generated using computer-aided drawing and design (CADD). Databases are often created using relational database software, such as ORACLE®. By activating information contained in a particular database, the map can tell an environmental "story" about the site.

Different "layers" of data are created in the database. Some of these layers, called *planimetrics*, can be environmentally oriented, for example, soil sampling data, groundwater elevations, and locations of ecological receptors. Other layers may be cultural in nature: the location of buildings and roads, for instance, or property boundaries. The user can select which layers will be plotted on the map. The GIS can be linked to software programs, such as automated contouring programs or programs that automatically generate cross-sectional views of geological strata. The flexibility offered by a GIS makes it a desirable method of data management, especially for complicated sites.

7.4 Vapor Intrusion Investigation

VOCs present in soils or groundwater underneath or near a building can volatilize and enter the building through cracks or openings in the building foundation, as shown in Figure 7.6. Known as *vapor intrusion*, this situation can present a health hazard to building occupants and workers. A relative latecomer to the pantheon of environmental concerns, it was not until the 1990s that a study by the Massachusetts Department of Environmental Protection brought the issue to the attention of the regulatory community. It is discussed in this chapter rather than Chapter 17 on indoor air quality because, although it is an indoor air quality issue, it is related to subsurface contamination and is often part of a remedial investigation.

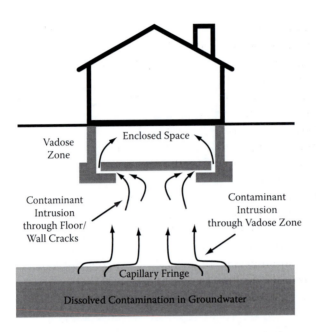

FIGURE 7.6
Conceptual diagram of vapor intrusion. (Adapted from the U.S. Environmental Protection Agency.)

7.4.1 Vapor Intrusion Investigation as Part of a Remedial Investigation

Usually vapor intrusion investigations are triggered when a remedial investigation indicates the presence of VOCs in soils or groundwater underneath or near a building. A vapor intrusion investigation can also be initiated when a material threat of vapor intrusion is identified in the course of a Phase I Environmental Site Assessment. Assessment of potential Vapor Encroachment Conditions (VECs) are increasingly becoming part of a Phase I ESA. Many states now require a vapor intrusion investigation when VOCs are known to exist within a certain distance from or on a target property, as discussed next.

7.4.2 Vapor Intrusion Investigation

A two-tier approach is often employed in assessing whether a potential vapor intrusion problem exists on a property. This approach was set forth in the ASTM Vapor Encroachment Screening standard (E2600-10), first published in 2008 and modified in 2010, which in turn was modeled after a draft guidance document published by the USEPA in 2002.

A Tier I screen is basically a due diligence approach similar in scope to the Phase I Environmental Site Assessment (see Chapter 5), and it uses many of the same terms. The objective of the Tier I screen is to determine if a VEC exists on the property. The goal is to attempt to identify whether nearby properties or the target property have chemicals that can create a VEC. The radius search for nearby properties covers 0.1 miles (0.16 kilometers) for petroleum-related VOCs and 0.33 mile (0.54 kilometers) for other VOCs. The E2600-10 standard allows these search distances to be decreased if the investigator has special knowledge about geologic conditions in the area or if the direction of groundwater flow is known.

Information sources for a Tier I screen are similar to the Phase I ESA information sources: interviews of knowledgeable parties; reviews of "reasonably ascertainable" and "readily available" environmental records available on the target property and at local and state agencies; historical record sources; general geological information regarding soil type, groundwater depth, and groundwater flow direction; and a standard database radius search. The radius search would help identify neighboring properties that would be of concern including known or suspected contaminated sites; dry cleaning establishments; gasoline stations; and industrial facilities that "are likely to" utilize solvents or petroleum products. A reconnaissance of the neighborhood would also help identify these "high-risk" properties.

A Tier I screen looks for potential pathways for vapor migration and considers the future use of the property. If the future use is unknown, then the investigator is required, under the E2600-10 standard, to make a conservative assumption, that is, that a sensitive population (residential, day care, etc.) will occupy the building.

The conclusions of the Tier I screen must also be conservative. If a VEC exists, is likely to exist, or *cannot be ruled out*, then a Tier II screen must be performed (called a Tier III assessment in the USEPA guidance document).

A Tier II screen involves at least noninvasive data collection. It includes review of site investigations and remedial investigations conducted on neighboring properties available in site or regulatory files to obtain a more thorough understanding of the threat posed by the neighboring property. After completion of the file reviews, the investigator must determine if there are petroleum-related chemicals within 30 feet (9 meters) of the building, or nonpetroleum related chemicals within 100 feet (30.5 meters) of the building. These distances, defined as the *critical distances,* can be reduced if there is special knowledge, usually relating to subsurface conditions, suggesting as such. If the possibility for a VOC hasn't then been eliminated, invasive sampling is warranted.

7.4.3 Vapor Sampling Procedures

There are two basic types of indoor air samples that are collected as part of an invasive vapor intrusion investigation: indoor air samples and sub-slab samples. Usually collected before indoor air samples are collected, *sub-slab sampling* (see Figure 7.7) involves drilling a narrow hole through the foundation floor of the building and collecting a sample of soil gas in a *summa*

FIGURE 7.7
A summa canister being used for sub-slab vapor sampling. (Courtesy of Dermody Consulting.)

canister. Ideally, the hole is drilled in the part of the building closest to the potential VOC source, although variations in building construction should also be taken into account when locating sub-slab sampling points. Once it is filled with soil gas, the summa canister is then sent to a laboratory for analysis, usually using USEPA Method TO-15.

Indoor air samples are collected by placing summa canisters on the lowest floor of the building at strategic locations, such as nearest the plume or near floor openings, such as sumps. A canister is also placed outside the building for quality control purposes. The canisters collect air for a given time period, usually 24 hours (although recent studies suggest that longer air collection periods can reduce temporal variations), and then are sent to a laboratory for analysis. Before collecting the samples, it is advisable to go through the standard steps in any indoor air quality investigation, as described in Chapter 17: taking an inventory of chemicals that are present inside the building; looking for potential chemical sources in the building, such as dry cleaning, new carpeting or other materials that can give off gasses; and so forth. Chemicals should be removed prior to collecting the air samples to the extent feasible, since these chemicals are likely to be detected in the indoor air sample.

Once the results of the laboratory analysis are obtained, they are either compared to results to standards set by the regulatory agency or guidelines, or to generic values that are calculated by making various conservative assumptions and then employing standard risk assessment calculations, which are discussed in Section 7.5.2. For a pathway to be complete, *the same chemical must be present in the subsurface, in the sub-slab sample, and in the indoor air sample.* Table 7.3 provides a reference by which to interpret vapor sampling results.

TABLE 7.3

Meaning of Vapor Intrusion Sampling Results

Is Chemical Present In			
Indoor Air Sample	Sub-Slab Sample	Subsurface	Interpretation
No	Yes	Yes	No vapor intrusion issue inside the building.
Yes	No	Yes	No vapor intrusion issue inside the building. However, an indoor air quality issue may exist. Check for other sources of indoor air pollution, such as chemical storage or usage inside the building.
Yes	Yes	No	An indoor air quality issue may exist, but it is not related to the known subsurface contamination. No vapor intrusion issue inside the building.
Yes	Yes	Yes	Contaminant is present in the subsurface, in soil gas just beneath the foundation, and in the indoor air. Pathway is complete. A vapor intrusion issue exists inside the building.

If the contaminant detected in the indoor air did not originate from the subsurface plume, then the investigator needs to determine its source. Going through the facility's chemical inventory may reveal that the chemical likely originated inside the facility rather than the subsurface. Interviewing the building owner and tenant may be advisable. Presence of the chemical in the sub-slab sample but not the subsurface plume might indicate "short circuiting" in the chemical pathway, which could occur when a chemical inside the building migrates through a preferential subsurface pathway and be picked up in the sub-slab sample. If there is nothing to suggest an on-site source, then an off-site contribution is suspected.

7.4.4 Vapor Intrusion Mitigation

If a vapor intrusion issue exists at the building, then a vapor mitigation system must be put into place. This system is basically a radon mitigation system, which is described in Chapter 16. There is a reason for this: radon gas emanates from a geological formation just like vapors emanating from subsurface contamination and works its way into the building using the same pathways.

Unlike radon, the source of the vapor intrusion problem can go away. Once the source plume has been remediated, the vapor mitigation system is no longer needed and can be turned off. Periodic indoor air and sub-slab sampling may also be warranted. Sub-slab sampling should be performed once the system has been turned off for a period of time to see if the pathway into the building is still complete.

7.5 Setting Remediation Standards

Setting remediation standards for a contaminated property may seem like a straightforward proposition—clean up everything that polluted the property! However, real-life conditions tend to be far more complex, since virtually no previously developed property in the United States is chemical free. Tiny quantities of chemicals have migrated onto otherwise pristine properties due to general human activities. Chemicals have settled onto the ground from the air, in some cases traveling thousands of miles from their point of origin. Even the soils and rocks themselves naturally contain chemicals that are toxic to humans, such as lead and arsenic. For this reason, an environmental investigation will often detect some quantity, however minute, of hazardous chemicals.

But does their mere presence mean the property requires remediation? Why not remove all contamination? Can some of the contamination remain, and the site be considered "clean"? Or, phrased another way, how clean is clean?

In a perfect world, polluted properties should be remediated to "pre-industrial" levels, a term used by the New York State Department of Environmental Conservation (NYSDEC). This ideal, however desirable, is impractical. Nor will it necessarily result in a better or healthier environment for humans, flora, and fauna. Removal of contaminated media can by itself create a health hazard, through the migration of contaminants into the air and water, where none previously existed. In the case of naturally occurring elements, such as lead or arsenic, the contamination cannot be removed completely. Even in cases where the contamination is caused by human processes, it is often physically impossible to remove all contamination.

Rather than remediate all properties to pre-industrial conditions, *remediation standards* are established to guide the remediation of contaminated media.

7.5.1 Generic Remediation Standards

Ideally, the remediation standards for each property will reflect the risk posed by the site contaminants to human health and the environment. However, developing site-specific remediation standards for each site is time-consuming, expensive, and probably unnecessary in many if not most cases due to similarities in geology and chemistry.

For this reason, most regulatory authorities have *generic remediation standards* for the compounds or analytes required to be tested for the regulation being enforced (Superfund, Clean Water Act, etc.). These standards are designed to protect human health and the environment under most circumstances. The standards are conservative in that they err on the side of caution by basing the risk posed by the chemical on ideal fate and transport conditions:

- Soils that have high porosity and permeability in the soil formations, enabling free movement of contaminants in the subsurface
- High rate of exposure to potential receptors; and
- Presence of sensitive receptors, that is, those receptors (human or ecological) most vulnerable to chemical exposure

Generic standards are based on a *dose–response* relationship, and are developed using the risk-based principles described in the next section. The dose-response concept assumes that the higher the dose of a certain chemical, the more likely a negative impact of the chemical. This principle was famously stated by Paracelsus, the father of toxicology: "the dose makes the poison."

Often, there are multiple generic cleanup standards to cover multiple generic scenarios. The most stringent generic cleanup standards for soils are reserved for residential properties, which are usually designed to protect the health of resident children. Nonresidential standards are less stringent than residential standards because the affected population, assumed to consist primarily of adults, has a larger body mass with which to absorb contaminants and is present on the site for less time (40 hours per week versus 24/7). Another commonly used generic standard for soils is designed to protect contaminants in soils from migrating downward to the water table where it can be ingested if the groundwater is used for consumption. These standards are generally known as "impact to groundwater" standards.

Generic groundwater standards include standards for drinking water, consistent with the Clean Water Act; less stringent generic standards for nonpotable or brackish groundwater; and the most stringent standards for protected watersheds and other specially designated water bodies. Surface water and sediment standards are discussed in Chapter 9.

7.5.2 Setting Cleanup Goals Using Risk Assessment

To estimate the threat to human health posed by contaminants on a specific property, and therefore understand the risk to human health posed by a specific hazard, a *risk assessment* is performed. Developed by the USEPA in response to a Superfund program mandate, the risk assessment is usually performed at or near the conclusion of the remedial investigation when the data set is complete or nearly complete, although, depending on the purpose, a risk assessment could be conducted at early stage of the investigation, for example, for screening and for assisting site investigations.

A risk assessment can be performed in two different "approaches." The most common approach of a risk assessment is to identify a potentially hazardous condition and then determine the risk to human health posed by that hazard. The opposite approach, utilized in risk assessments associated with remediation, is to identify a hazard (e.g., contaminated soil), select a level of *acceptable risk* posed by that hazard, and then back calculate the level of contamination that is "acceptable" given the risk toleration.

A *baseline risk assessment* is an analysis of the potential adverse health effects (current or future) caused by hazardous substance releases from a site. It assumes that no actions will be taken to control or mitigate these releases, and therefore focuses on current site conditions rather than conditions that may exist postremediation or postconstruction. The results of the baseline risk assessment will help determine what remedial actions should be implemented at a site by developing site-specific remediation goals and documenting the magnitude of risk posed by a site. Alternatively, a *focused risk assessment* analyzes the risk from a particular chemical or a subset of the chemicals present on a site.

The baseline risk assessment consists of five steps:

1. Hazard identification
2. Data collection and evaluation
3. Exposure assessment
4. Toxicity assessment
5. Risk characterization

The first two steps are largely accomplished by the Phase I/site investigation process described in this and the previous two chapters. The data collected for these two steps include not only the concentrations of the contaminants on the site and their spatial distribution, but also their expected fate and transport and potential receptors. Steps 3 and 4, the exposure assessment and toxicity assessment, can be performed concurrently, concluding with the risk characterization, all three of which are described next. The focused risk assessment usually starts at steps 3 and 4, with the particular chemical that is the subject of the focused assessment already identified and evaluated.

7.5.2.1 Exposure Assessment

Identifying pathways to potential receptors, and the estimated dosage to the receptors, is known as an *exposure assessment*. It provides an estimation of the magnitude, frequency, duration, and route of exposure for a given contaminant. More specifically, it identifies potentially exposed populations (read: receptors) and potential pathways, and estimates exposure concentrations and chemical intakes. Since the baseline risk assessment assumes that no remedial actions will be performed, it must take into account currrent as well as future pathways and receptors.

Typical human receptors to consider in an industrial setting include:

- Facility workers who may have daily contact with contaminated media

- Utility emergency repair workers who may come into contact with contaminated soils when dealing with underground utilities
- Construction workers who may be doing maintenance and repair, or may be constructing new buildings on the property
- Trespassers/authorized site visitors

Potential receptors in a residential setting also include sensitive receptors such as children.

A *human health risk assessment* focuses on the inhalation, ingestion, and dermal exposure pathways into the human body. (Environmental risk assessments focus on other biota.) Included in the inhalation pathway are the inhalation of contaminants in the gaseous phase as well as contaminants attached to dust particles. The inhalation pathway also considers inhalation of contaminated water by a human who is in direct contact with the water, such as with showering or bathing. The ingestion pathway includes drinking contaminated water or eating food that has absorbed contaminants present in the soil. Children can also ingest contaminants by playing outside in contaminanted soils and placing their dirty hands in their mouths. Table 7.4 provides exposure scenarios used by the NYSDEC in establishing its generic

TABLE 7.4

Exposure Scenario Receptors and Pathways

Land Use Category	Unrestricted	Residential	Restricted Residential	Commercial	Industrial
Exposed Population	Adult and Child	Adult and Child	Adult and Child	Adult and Child	Adult and Adolescent
Route of Exposure					
Incidental soil ingestion	√	√	√	√	√
Inhalation of soil	√	√	√	√	√
Dermal contact with soil	√	√	√	√	√
Homegrown vegetable consumption	√	√			
Producing animal products for human consumption	√				
Groundwater protection	√	√	√	√	√
Ecological resource protection	√	√	√	√	√

Source: Adapted from New York State Department of Environmental Conservation, May 2010, DER-10: Technical Guidance for Site Investigation and Remediation.

cleanup standards and summarizes the prime exposure pathways for a human health risk assessment.

Each exposure route used in a human health risk assessment utilizes a different equation in calculating the *reasonable maximum exposure* (RME). The RME is the highest exposure that could reasonably be expected to occur for a given exposure pathway at a site. It should account for both uncertainty in the contaminant concentration and variability in the exposure parameters (such as exposure frequency and averaging time).

For instance, the equation used to calculate exposure from inhaling airborne dust for an adult in an industrial setting is

$$\text{Intake (mg/kg/day)} = (EPC_{air} \times IR \times EF \times ED)/(BW \times AT)$$

where

EPC_{air} = exposure point concentration in air (mg/m^3)
IR   = inhalation rate (m^3/day)
EF   = exposure frequency (days/year)
ED   = exposure duration (years)
BW   = body weight (kg)
AT   = averaging time (days)

In this equation, the EPC_{air} is determined from the exposure assessment. To obtain an inhalation rate for this equation, the USEPA estimates that the average adult inhales 20 m^3 of air per day. This is a conservative assumption, as are the exposure frequency to be considered in this exposure scenario (250 days per year for workers in industrial settings, 365 days per year for residents in residential settings) and the exposure duration (25-year careers for workers in industrial settings, 70-year residency for people in residential settings). The "standard adult" considered in the risk assessment weighs 70 kg (approximately 154 lbs). The body weight is in the numerator, because in theory people with greater body weight can tolerate a greater quantity of toxins. The averaging time for noncarcinogens is the period of concern for the study. Similar calculations are performed for the other exposure paths.

7.5.2.2 Toxicity Assessment

A *toxicity assessment* entails the acquisition and evaluation of toxicity data for each contaminant present at a Site. It involves research into the latest published values to ensure that the most recent research involving the chemicals of concern is employed in the risk assessment process. Many states publish their own toxicity values for chemicals.

Toxicity assessments separate chemicals into two categories: (1) carcinogens, defined as chemicals with the ability to cause cancer in humans; and (2) noncarcinogens. The USEPA defines five groups regarding carcinogenicity in humans (www.epa.gov/ttn/atw/toxsource/carcinogens.html):

Group A—Carcinogenic to humans

Group B—"Probably" carcinogenic to humans

Group C—"Possibly" carcinogenic to humans

Group D—Not classifiable as to human carcinogenicity

Group E—Evidence of noncarcinogenicity for humans

In a toxicity assessment, the category of carcinogens includes Groups A and B only. The USEPA's Integrated Risk Information System (IRIS) Web site (www.epa.gov/iris/search_human.htm) lists the chemicals for Groups A, B, and C.

Carcinogenic risk for a chemical is estimated using a *slope factor*, which is an upper bound, approximating a 95% confidence limit, on the increased cancer risk from a lifetime exposure to a chemical by ingestion or inhalation (see Figure 7.8). Superfund defines an "acceptable risk" that exposure to a chemical will cause an incremental cancer (in addition to the 1-in-4 chance) at 1×10^{-6}, or one person per million population. The 1×10^{-6} risk is used in a risk assessment as the "acceptable risk" to carcinogens present at a hazardous waste site.

The risk posed by noncarcinogens at a contaminated site is estimated using *reference doses* (RfDs) for the ingestion and dermal exposure pathways and *reference concentrations* (RfCs) for the inhalation pathway. Reference doses, which are values supplied by the USEPA, assume that for noncarcinogens there is a dosage below which there is no health effect from human exposure to a chemical. Above this dosage, known as the *no observed adverse effect level* (NOAEL), there is an assumed linear relationship between dosage and health effect (see Figure 7.9). The RfD is equal to the NOAEL multiplied by a safety factor, in many cases a factor of 10. The RfC is derived in a similar manner to the RfD, except that several inhalation exposure scenarios are considered,

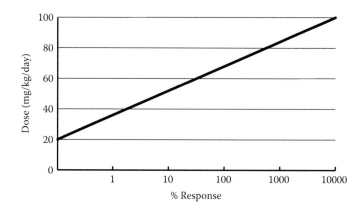

FIGURE 7.8

A graph showing the 95% confidence interval for a known or probable carcinogen.

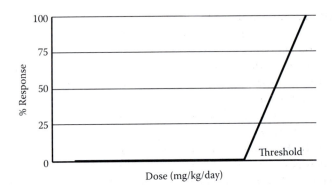

FIGURE 7.9
A graph showing the relationship between dosage and health effect for a noncarcinogen.

with the most toxic scenario for a given chemical used in the toxicity assessment. USEPA has set a noncarcinogenic hazard effect target limit value of 1.

The USEPA recommends the following hierarchy of toxicological sources for Superfund Risk Assessments (OSWER Directive 9285.7-53, December 5, 2003):

Tier 1—IRIS.

Tier 2—USEPA's provisional peer reviewed toxicity values (PPRTVs)— The Office of Research and Development/National Center for Environmental Assessment/Superfund Health Risk Technical Support Center develops PPRTVs on a chemical-specific basis. These values are available upon request through the project USEPA risk assessor for Superfund projects.

Tier 3—Includes additional USEPA and non-USEPA sources, such as the California EPA toxicity values (www.oehha.ca.gov/risk/ChemicalDB) and the Agency for Toxic Substances and Disease Registry (ATSDR) Minimal Risk Levels (www.atsdr.cdc.gov/mrls.html).

Since many dozens of chemicals are often present at a hazardous waste site, it is often impractical to provide a toxicity assessment for each chemical present. Instead, chemicals that are chemically similar to other chemicals are chosen to represent its class. Such chemicals, known as *surrogate chemicals*, should be the most toxic, persistent, and mobile of the chemicals within its chemical class, and the most prevalent at the site. In cases where all of these criteria cannot be met, the risk assessor should either select one chemical that fits most of these criteria or select more than one chemical from the chemical class.

7.5.2.3 Risk Characterization

Once the aforementioned steps have been completed, the overall site risk can be calculated. Where multiple contaminants are present, the cumulative risks posed by those contaminants must be calculated, since estimating the risk posed by one contaminant at a time will underestimate the risk posed by a site. The risks by pathway for current and future scenarios are summarized, resulting in a *cancer risk* for that pathway posed by the site and a cumulative *noncancer hazard index* for the noncarcinogen contaminants at the site. The calculated risks then would be compared to the risk limits for risk evaluation conclusions.

For instance, the exposure and toxicity assessments may indicate that the overall cancer risk presented by a site is 7×10^{-7} and noncancer hazard index of 0.3. Since the overall cancer risk is below 1×10^{-6} and the noncancer hazard index is below the noncarcinogenic hazard effect target limit value of 1, one could conclude that the presence of the detected chemicals on the property do not pose an unacceptable health risk to the populations evaluated for the risk assessment.

A proper risk characterization will incorporate all major assumptions, scientific judgments, and estimates of the uncertainties present in the data. Once an uncertainty analysis is performed and all site-specific health or exposure studies taken into account, site-specific cleanup standards can be calculated, designed to protect human health within the assumed risk tolerance.

References

ASTM International, 2010. Standard Guide for Vapor Encroachment Screening on Property Involved in Real Estate Transactions. E-2600-10.

Freeze, R. Allan, and Cherry, John A. 1979. *Groundwater*. New York: Prentice Hall.

Guidotti, T. L. 1988. Exposure to hazard and individual risk: When occupational medicine gets personal. *Journal of Occupational Medicine* 30: 570–577.

Hemond, Harold F., and Fechner-Levy, Elizabeth J. 2000. *Chemical Fate and Transport in the Environment*, 2nd ed. Salt Lake City, UT: Academic Press.

Interstate Technology & Regulatory Council. January 2007. Vapor Intrusion Pathway: A Practical Guideline.

Lagrega, Michael D., Buckingham, Phillip L., and Evans, Jeffrey E. 2001. *Hazardous Waste Management*, 2nd ed. New York: McGraw-Hill.

National Oceanic and Atmospheric Administration. National Geodetic Survey: Frequently Asked Questions. http://www.ngs.noaa.gov/faq.shtml#WhatDatum

New Jersey Department of Environmental Protection. August 2005. Vapor Intrusion Guidance.

New York State Department of Environmental Conservation. May 2010. DER-10: Technical Guidance for Site Investigation and Remediation.

Nyer, Evan K., Regan, Terry, and Nautiyal, Deepak. 1996. Developing a healthy disrespect for numbers. *Ground Water Monitoring Review*, 59–64.

Nyer, Evan K., and Gearhart, Mary J. 1997. Plumes Don't Move. Groundwater Water Monitoring and Remediation. National Ground Water Association.

Patton, Dorothy E. 1993. The ABCs of risk assessment. *EPA Journal* 19: 10–15.

Sterrett, Robert J. 2007. *Groundwater and Wells*, 3rd ed. Johnson Screens.

U.S. Environmental Protection Agency. August 1987. A Compendium of Superfund Field Operations Methods. EPA 540/P-87/001. Office of Emergency and Remedial Response.

U.S. Environmental Protection Agency. 1989. Exposure Factors Handbook. EPA/600/8-89/043. Office of Emergency and Remedial Response.

U.S. Environmental Protection Agency. December 1989. Risk Assessment Guidance for Superfund, Volume I. Human Health Evaluation Manual. EPA 540/1-89/002. Office of Emergency and Remedial Response.

U.S. Environmental Protection Agency. November 2002. Draft OSWER Guidance for Evaluating the Vapor Intrusion to Indoor Air Pathway from Groundwater and Soils (Subsurface Vapor Intrusion Guidance). Office of Solid Waste and Emergency Response. EPA-530-D-02-004.

U.S. Environmental Protection Agency. December 2002. Supplemental Guidance for Developing Soil Screening Levels for Superfund Sites. OSWER 9355.4-24. Office of Emergency and Remedial Response.

U.S. Environmental Protection Agency. December 2003. Hierarchy of Toxicological Sources for Superfund Risk Assessments. OSWER Directive 9285.7-53.

U.S. Environmental Protection Agency. 2006. National Recommended Water Quality Criteria. Office of Science and Technology (4304T).

U.S. Environmental Protection Agency. 2010. Review of the Draft 2002 Subsurface Vapor Intrusion Guidance. Office of Solid Waste and Emergency Response.

Weiner, Scott A. 1993. Developing Cleanup Standards for Contaminated Soil, Sediment, & Groundwater: How Clean Is Clean? Specialty Conference Series, January 10–13. Federal Water Environment.

8

Remedial Actions

As suggested in the previous chapters, the remediation of contaminated media is the culmination of an often laborious and time-consuming process of identifying and evaluating areas of concern; developing and refining the conceptual site model; obtaining hydrogeological, geochemical, physical, and biological data; and, not to be overlooked, documenting the observations and findings of the investigations.

This chapter discusses the basis for selecting remedial action and describes many but not all of the myriad of remedial actions available to the environmental consultant.

8.1 Remedial Action Selection Criteria

To design a remedial action, one must establish the *design basis* for the action. The design basis identifies the chemicals to be remediated and the remedial goals for each chemical.

8.1.1 Establishing Remedial Objectives

The design basis should conform to the remedial objectives. The endpoint of the remediation may call for unrestricted future usage, or future usage that does not allow for human occupancy or restricted, nonresidential occupancy. The remediation may be specific to an area of concern or it may be sitewide. Alternatively, where immediate health concerns or impacts to the environment are present, it may be more appropriate to employ interim remedial measures (IRMs) now rather than take the time to devise and implement a comprehensive remediation.

The site and its contaminants will usually place constraints on the selection of the remedial action; the selected remedial action must be implementable. The remedial action also will reflect a balancing of time and cost considerations. As a general rule of thumb, remedial actions that can be performed quickly will usually be more expensive and vice versa.

8.1.2 Feasibility Study

The process for selecting the remedial action is known as the *feasibility study*. The feasibility study involves the collection of data that will be used to design the remedial action and the evaluation of the data in light of the established remedial objectives. A list of remedial alternatives is developed and screened based on remedial action objectives as described earlier. Since it would take far too long to consider every remedial technology at every site, a representative technology process option for each technology type is usually selected. Estimated costs and timeframes are generated and used in the decision-making process.

In some cases, the feasibility study may include testing of various remedial actions in actual or simulated site conditions. Feasibility studies include *treatability studies* (also known as bench-scale tests) and *pilot tests*.

A treatability study involves laboratory simulations of site conditions. The study may involve the site contaminants in proportion to their site conditions or entail isolating one site contaminant, and use actual site soils to observe the effectiveness of the remedial technology given expected chemical, physical, and biological conditions. A *pilot test* implements a prospective remedial method over a small portion of the contaminated area. The selected area should be representative of conditions across the entire contaminated area. If conditions vary across the area, more than one area should be tested or more than one pilot test performed if multiple remedial methods may be employed. The pilot test has the advantage of being easily scaled up to a full-scale remediation.

8.2 Types of Remedial Actions

Remedial actions broadly fall into two categories: passive remediation and active remediation (although some remedial technologies blur the distinction between the two categories). *Passive remediation* entails constructing engineering controls to control the migration of and exposure to contaminants or the establishment of institutional controls to prevent human contact with the contaminants. *Active remediation* involves applying a physical, chemical, or biological process to the contaminants in an effort to decrease their concentration to acceptable levels. It can occur *ex situ* (on the ground surface) or *in situ* (in the subsurface). The passive and active remedial methods most commonly employed at contaminated sites are described in the following subsections.

8.2.1 Passive Remediation and Engineering Controls

Engineering controls are designed and constructed with a specific purpose: to prevent contaminants from entering the human body or the environment.

Engineering controls do not attempt to reduce or eliminate the source of the contamination.

The most common type of engineering control for the dermal pathway is an impermeable surface cap. This cap may consist of asphalt or concrete, or impermeable soil (such as clay) of sufficient thickness as to reliably prevent future direct human contact with the contaminated soils. The barrier can also be a manufactured barrier known as a *geosynthetic liner*, which is an artificial barrier usually comprised of high-density polyethylene (HDPE) or equivalent. These liners are commonly used in the permanent closure of solid waste landfills (see Section 8.3). A surface cap is highly preferable to passive engineering controls, such as fencing around the contaminated area, since fences and other such controls are relatively easy to circumvent. However, caps are typically far more expensive than more passive engineering controls, so a cost-benefit analysis may need to be performed.

Another type of engineering control is the *vertical containment barrier*. This barrier, which is usually a concrete slurry wall or steel sheet piling, is designed to prevent the migration of contaminants toward a receptor, such as a drinking water well or a surface water body. Like surface caps, vertical containment barriers do not reduce or eliminate the contaminant source; they address the pathway into the human body or the environment only.

Institutional controls involve effecting a legal deterrent to human activities in the contaminated area. Often combined with engineering controls, they are sometimes known as *activity and use limitations* (AULs). The principal mechanism of an institutional control is the *deed notice*, which is attached to the property deed and indicates the location and nature of the contaminated area. Deed notices usually restrict human activity in this area. AULs may also include the reclassification of the groundwater or the surface water body in the contaminated area, or restrictions on its future use as a water source for drinking or irrigation.

If implemented properly, engineering and institutional controls should be protective of human health and the environment. However, these controls are only as good as the people who must abide by them in the future. As discussed in Chapter 2, engineering and institutional controls did not prevent the Love Canal disaster.

8.2.2 Monitored Natural Attenuation

Often confused with AULs is *monitored natural attenuation* (MNA), which involves the reduction in the toxicity of dissolved contaminants by natural degradation. MNA is employed for contaminants that will degrade naturally, either by naturally occurring chemical or biological reactions. These reactions may involve oxidation or reduction, depending on the contaminant and the natural setting, as described later in this chapter.

To utilize MNA as the remedial action, it first must be established that the contaminant(s) will degrade naturally. This may be done through field studies or in a bench scale study in the laboratory. MNA follows the patterns described later in this chapter under *in situ* remediation. For MNA to occur, moisture must be available (even in the vadose zone), since the contaminants must be in the dissolved phase for the reactions to occur.

Reactions may be oxidative or reductive. In general, biotransformation, where it is the dominant process, occurs more quickly by aerobic microorganisms rather than anaerobic microorganisms. For this reason, natural aerobic biodegradation works best when abundant oxygen is present, which in general is more common in shallower soils and groundwater.

Once MNA is implemented, it will be necessary to periodically check indicators that MNA is occurring: contaminant concentrations, byproducts of the expected geochemical or biochemical processes, and levels of the indigenous microbes that are expected to be causing the biotransformations. Since the process usually occurs slowly, it is often coupled with the establishment of a groundwater use restriction (an institutional control) in the contaminated area.

It should be noted that the U.S. Environmental Protection Agency (USEPA) and most other regulatory authorities do not accept dispersion and dilution as MNA processes. In other words, "The Solution To Pollution Is Not Dilution!"

8.2.3 Removal of Nonaqueous Phase Liquid (NAPL)

NAPL, also known as free product, is an ongoing source of contamination, leaching a constant stream of contaminants into the soils and groundwater. Until all NAPL is removed, remediation of dissolved constituents is impossible to achieve, since contaminants will continue to dissolve from the NAPL into the available moisture. Another reason for considering NAPL separately from the remediation of dissolved constituents is its resistance to many of the *in situ* remediation methods discussed later.

Most cases of NAPL involve LNAPL (light nonaqueous phase liquid), and most LNAPL is a petroleum product. Fortunately (and predictably), LNAPL floats on top of the water table, since it is by definition lighter than water. Due to this convenience, most technologies for the remediation of LNAPL involve physical removal. Some popular types of LNAPL removal are described next.

The simplest method of passive LNAPL remediation is known as *skimming*, in which withdrawal points (with similar construction to monitoring wells) are installed within the LNAPL-impacted area. LNAPL can be manually removed from these withdrawal points using specially designed bailers or with pumps that are activated when they encounter a liquid with high resistivity (as opposed to water, which has a much lower resistivity than petroleum). In some cases, interceptor trenches are constructed through or downgradient of the impacted area below the depth of the LNAPL. The LNAPL naturally flows into the trench, where it is then recovered using a pump or some manual means.

Bioslurping is a vacuum-enhanced method of LNAPL recovery. In bioslurping, a seal is created on an extraction well and a "bioslurp" tube is installed within the LNAPL. A vacuum is applied to the tube, which acts like a straw in sucking the LNAPL to the surface. It also introduces air into the subsurface, encouraging the growth of microorganisms in the same manner as bioventing, which is discussed in Section 8.2.6.8.

Of course, there is always the brute force method of LNAPL removal, in which a pump is used to remove all liquids, LNAPL as well as water, from the contaminated area. This method will result in the generation of a large quantity of wastewater, which could result in excessive wastewater disposal costs. However, that wastewater can be sent through an oil–water separator and reintroduced into the geologic formation once free of petroleum, thereby reducing the volume of wastewater and its related disposal costs. Pumping also has the advantage of depressing the water table, which will create a cone of depression toward which water and LNAPL will flow, thus enhancing the recovery of the LNAPL in the contaminated area.

All of the aforementioned methods can be combined with various *in situ* (in-place) technologies, such as the injection of surfactants into the subsurface (see Section 8.2.6.3), which free the LNAPL from the soil particles, facilitating their movement and recovery, or the injection of a cosolvent to liberate the LNAPL from soils and other solids, and enable it to flow toward the recovery location.

DNAPL remediation is generally performed by removal methods, such as groundwater pump-and-treat (see Section 8.2.5). There are emerging technologies designed to remediate DNAPL. More information is available on USEPA's Clu-In Web site (www.clu-in.org).

8.2.4 Soil Remediation by Excavation

The technologically simplest way to remediate contaminated soils is by digging them up and sending them off site for treatment or disposal. Excavating equipment can be large, as shown in Figure 8.1, or can be small as a golf cart or even a hand shovel. After excavating the target area, the completeness of the remedial action should be confirmed by the collection of soil samples, known as *post-excavation samples*. Post-excavation samples should be collected from the sidewalls and bottom of the excavation in a sufficient quantity to confirm the uncontaminated nature of the remaining soils. The samples should be analyzed for the contaminants that had been present in the soils that were remediated. The excavation can be backfilled with clean fill or soils, or the treated soils once it has been verified that the post-excavation soil samples have acceptable concentrations of the target contaminants.

Disposal of the contaminated soils must comply with the Resource Conservation and Recovery Act (RCRA), Toxic Substances Control Act (TSCA), and all applicable state and local laws. Unless the soils are listed

FIGURE 8.1
An excavator loading contaminated soils onto a dump truck. (Courtesy of GZA GeoEnvironmental, Inc.)

wastes (see Chapter 2), the waste soils must undergo a series of analyses to determine their *waste classification* to comply with RCRA and TSCA land disposal regulations. This waste classification is performed prior to shipment of the soils to a waste receiving facility, since there are severe penalties for disposal facilities that do not follow the RCRA procedures for waste acceptance. It includes tests for RCRA characteristics (ignitability, corrosivity, and reactivity), TCLP analyses of the compounds listed in Table 2.2, and analysis of total PCBs for TSCA compliance.

If the contaminated soils do not test hazardous, they cannot be considered to be "clean" and left in place. If the concentrations of regulated chemicals exceed standards set in accordance with the Clean Water Act, the Safe Drinking Water Act, or some other applicable statute, then they must be either treated to decrease their toxicity or removed. These soils generally are designated as contaminated, nonhazardous soils, and can be disposed of at landfills that are licensed to accept contaminated soils or otherwise treated and rendered nontoxic.

8.2.5 Groundwater Remediation by Pump-and-Treat

The analog to soil excavation for groundwater is known as *pump-and-treat*. As its name implies, pump-and-treat involves pumping contaminated groundwater to the surface and removing the contaminants from the groundwater. In Figure 8.2, contaminated groundwater is pumped to the surface through an extraction well and piped into a wastewater treatment system. Once separated from its contaminants, the treated groundwater is then injected into the subsurface through an injection well, while the removed contaminants are sent into the vapor phase and treated prior to venting them to the atmosphere. Pump-and-treat systems may also return the treated groundwater to the ground through an injection well or pipe it to a surface water body or sewer system. Other treatment options for the removed contamination are discussed next.

If the contaminant is inorganic, then it can be removed from the groundwater by means of *floccuation*. A common practice in the wastewater treatment industry, floccuation involves the addition of a chemical additive to the water stream that will form insoluble solids, such as hydroxides, sulfides, or carbonates, that precipitate from the water as solids. The solids are then

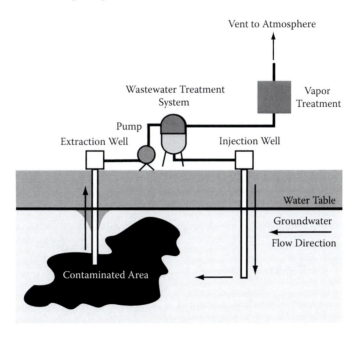

FIGURE 8.2
A schematic diagram of a typical pump-and-treat groundwater system.

recovered and diposed of in a proper manner. Organic compounds can be captured by running the groundwater through *granular activated carbon* (GAC). GAC is a form of carbon with a large surface area, which allows it to adsorb a wide range of compounds.

When the contaminants are volatile organic compounds (VOCs), they may be released in gaseous form to the atmosphere with or without treatment, depending on the quantity of contaminants and local regulations. If the regulatory or technical situation does not allow VOC venting to the atmosphere, then the VOCs must be captured once removed from the groundwater. This can be accomplished using GAC or a process called *air stripping*. Air stripping involves removing VOCs from the groundwater by aerating the groundwater, which will result in volatilizing the VOCs. Aeration can occur in a "packed tower," which is a device in which the contaminated groundwater is sprayed through a nozzle over packing inside the tower, and a fan blows air upward into the sprayed groundwater. The decontaminated groundwater is collected from a sump at the bottom of the tower.

Once removed from the groundwater, VOCs can be destroyed by biological or chemical processes, or captured using GAC. Another method of destroying captured VOCs is the process of ultraviolet oxidation, in which the contaminated groundwater is treated with oxygen, typically in the form of ozone or hydrogen peroxide, and ultraviolet rays. The energy from the ultraviolet rays spurs oxidation of the contaminants, resulting in their destruction. This process is effective with simple aromatic compounds, such as BTEX (benzene, toluene, ethylbenzene, and xylenes), as well as ethenes, such as perchloroethylene (PCE) and trichloroethylene (TCE).

To monitor the operation of a pump-and-treat system, a number of tests are typically performed. To confirm that the pumping wells have the proper drawdown and therefore the expected capture zone, groundwater level measurements periodically are collected from the monitoring wells within the zone of influence. Monitoring the flow rates within the extraction wells is important, since buildup of calcium, iron, and other fouling agents can reduce groundwater inflow into the extraction well and therefore reduce pumping rates and system effectiveness. Collecting periodic samples of the treated effluent indicates the effectiveness of the treatment portion of the pump-and-treat system.

Periodic sampling of the wells within the groundwater plume will document the effectiveness of the pump-and-treat system in reducing contaminant levels within the plume. As often occurs in pump-and-treat systems, there comes a point where the contaminant concentrations become stable, or *asymptotic*, after a periodic of decreasing concentration. When this occurs, contaminants hidden in the nooks and crannies of the geologic formation balance the ongoing mass removal of contaminants, resulting in no net change in concentration. In such cases, the pump-and-treat system has reached a point of diminishing returns and must either be turned off for a time period to allow the contaminants in the subsurface to equilibrate, be reconfigured, or be shut down completely.

One of the main drawbacks of pump-and-treat of groundwater is the enormous amount of effort and cost it takes to handle the water. Consider an aquifer that contains a VOC, say benzene, at a concentration of 1 milligram per liter (mg/l), which has a federal maximum contaminant level (MCL) of 0.005 mg/l. Of all of the liquids that are treated in the pump-and-treat system, 0.9999% of each milligram is just water. Pumping groundwater to the surface, moving it through a treatment system, and discharging it requires huge amounts of energy, with associated huge costs. A tiny fraction of this cost involves the contaminants themselves.

Pump-and-treat is not only a very expensive method to treat contaminated groundwater, but one that is, in many if not most cases, doomed to failure. As a general rule, the more complex the site's geology and chemistry, the more likely the pump-and-treat system will fail to achieve its cleanup goals (National Research Council, 1994). Among the many reasons for the failure of pump-and-treat systems in achieving groundwater quality standards in contaminated water are the following:

- Heterogeneities in the geologic formation—Heterogeneities can result in the creation of preferential groundwater flow patterns, bypassing contaminants located in less permeable formations.
- Migration of contaminants into inaccessible portions of the aquifer—This, too, will result in the bypassing of contaminants in the groundwater withdrawal process.
- Sorption of contaminants onto soil and rock particles—Pump-and-treat does not break the ionic bonds between contaminant and particle, which leaves behind large amounts of contaminants in certain geologic formations.

Several remedial technologies are designed to address the sorption of contaminants onto soil and rock particles in a pump-and-treat system. *Steam-enhanced extraction* involves the introduction of steam through injection wells. By heating the contaminants, they can volatilize into the gaseous phase (if VOCs), enabling them to be extracted from the subsurface through hot water recovery wells. The steam also helps mobilize the sorbed contaminants, enabling them to be removed from the subsurface.

Sorption of contaminants can also be addressed by injecting surfactants into the geologic formation rather than steam, which is expensive to generate and can easily cool down and lose its effectiveness in the subsurface. The surfactant, typically a specially designed detergent, performs the same function as detergent in your washing machine—it lowers the interfacial tension between the contaminants and water, and enables them to be removed from the media, be it subsurface soils or your clothing.

In recent years, an increasingly common usage of the groundwater pump-and-treat system has been the containment of the contaminant plume. Pumping groundwater from within or upgradient of the contaminant plume can slow or prevent the downgradient migration of the plume. Obtaining physical characteristics of the soils and the aquifer will assist the geologist in determining the minimum pumping rate needed to retard or stop the plume's migration.

8.2.6 *In Situ* Remediation of Dissolved Contaminants

Due to the high costs associated with soil excavation and pump-and-treat systems, the frequent ineffectiveness of pump-and-treat systems in accomplishing their objectives, and the complications involved in soil remediation when there are physical obstructions such as buildings or underground utilities, numerous technologies were developed, starting in the 1980s, involving *in situ* remediation.

VOCs are less amenable to *ex situ* remediation because their low molecular weight and low partition coefficients tend to make them more mobile than other contaminants and their plumes more widespread, both horizontally and vertically, than other contaminant plumes. Conversely, these same physical characteristics make them more amenable to *in situ* remediation methods, since they can be mobilized and vaporized more easily than other contaminant classes.

Several *in situ* technologies that are frequently employed at sites are discussed next. Summaries of these methods and other *in situ* methods are available at USEPA's http://clu-in.org/remediation/.

8.2.6.1 *Soil Vapor Extraction*

Soil vapor extraction (SVE) remediates VOCs in the unsaturated zone by taking advantage of their volatility. An SVE system induces air flow through the contaminanted portion of the subsurface and makes use of the ideal gas law to effect remediation (see Figure 8.3).

A map plan of an SVE system that might be employed to remediate the TCE in the soils at our fictitious manufacturing facility is shown in Figure 8.4. Boreholes are installed upgradient and downgradient of the contaminated zone. The boreholes installed upgradient of the contaminated zone are converted into injection wells, which are screened across the zone of interest. A compressor pushes air through the injection wells and into the geologic formation. On the downgradient side of the contaminated zone are vapor extraction wells, also screened across the zone of interest. Vacuum pumps attached to the vapor extraction wells remove air from the geologic formation. The combination of push and pull of the air creates air flow across the formation, and the partial vacuum created induces VOCs to go into the vapor phase.

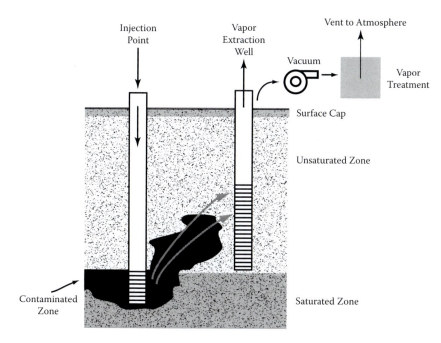

FIGURE 8.3
A schematic diagram of a soil vapor extraction (SVE) system.

Once in the vapor phase, the VOCs are removed from the contaminated zone through the vapor extraction wells.

Construction of the SVE system involves the following steps:

1. Installation of the injection wells and the vapor extraction wells.
2. Construction of trenches and the placement of underground piping, which connect the injection wells and the vapor extraction wells to the SVE equipment.
3. Installation of the SVE equipment (compressors, vacuum pumps, etc.). The SVE equipment is often located in one unit, which is sometimes brought to the site preassembled. Some type of capture mechanism is usually included in the SVE system, such as GAC described earlier, unless the VOCs can be vented to the atmosphere untreated.

In addition, an impermeable cap or low permeability cap or formation should be present on or near the surface for the SVE system to work effectively. Asphalt or concrete-paved surfaces form effective caps, as do overlying clay layers.

Short-circuiting of the air flow pattern is a system flaw that can impact the effectiveness of the SVE system. As the word implies, short-circuiting occurs

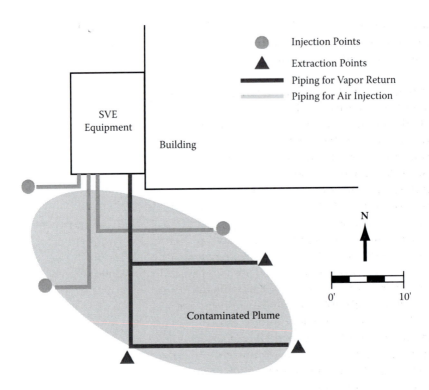

FIGURE 8.4
A schematic map plan of an SVE system at a fictitious manufacturing facility.

when air flow bypasses the zone of interest by finding a preferential pathway in the subsurface or due to the inadvertent creation of a preferential pathway to the surface (if the surface cap is imperfect or damaged).

Monitoring of an SVE system involves periodic checking on the pressure increase at the injection wells and the pressure decrease at the vapor extraction wells. Samples of the removed air may be collected to confirm the removal of contaminants from the geologic formation and the absence of short-circuiting in the subsurface.

As with many *in situ* remediation techniques, favorable geologic and chemical conditions must be present for the technique to work. The soils must have adequate permeability to allow for proper air movement (also there should not be isolated zones of low permeability—the air flow regime in the SVE could selectively bypass such zones, leaving them unremediated). The soils should lack physical properties that tend to retard the movement of VOCs, such as high organic carbon content. The contaminant of concern should easily go into gaseous phase, making the method less amenable to the heavier VOCs, or VOCs with a boiling point higher than 150°C (Hutzler et al., 1989).

8.2.6.2 Air Sparging

An *air sparging* (AS) system is similar in concept to an SVE system, except that it is designed to remediate the saturated zone (see Figure 8.5). Where VOC contamination is present in both the unsaturated and saturated zones, dual vacuum extraction systems, consisting of both SVE and AS components, are often employed, at a cost savings to the project.

In an air sparging system, injection wells, installed upgradient of the contaminated area, are screened across the contaminated portion of the saturated zone rather than the vadose zone as with an SVE system. The injection of air through the injection wells creates bubbling in the groundwater, sending VOCs into the vapor phase. Once in the vapor phase, the VOCs are removed from the formation through the vapor extraction wells, which are located downgradient of the contaminated area. Off-gas treatment, if performed, is similar to that described for SVE systems.

Construction and monitoring of an AS system is similar to the construction and monitoring of an SVE system. It should be noted that the same concerns regarding short-circuiting of the air flow regime applies to AS systems as well as SVE systems.

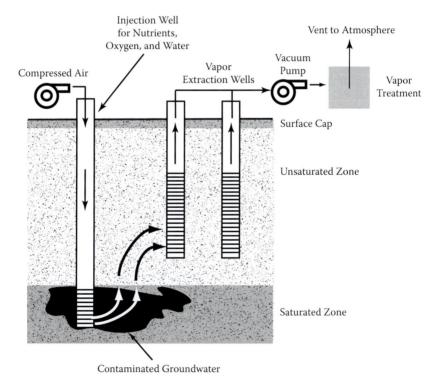

FIGURE 8.5
A schematic diagram of an air sparging (AS) system.

8.2.6.3 Soil Flushing

Soil flushing involves the introduction of surfactants into the saturated zone to liberate sorbed contaminants. In this process, surfactants are injected into the saturated zone through injection wells located upgradient of the contaminant plume, as described earlier in this chapter in conjunction with groundwater pump-and-treat systems and LNAPL recovery systems. The surfactants mobilize the contaminants, which are then recovered in recovery wells, along with the surfactant, which can be distilled and reused.

Sometimes cosolvents, which are substances that enhance the solubility of some organic compounds in water, also are injected to enhance soil flushing. Colsolvents may be organic compounds, such as alcohols, or ketones, such as acetone. However, some of these cosolvents are contaminants themselves, so they must be used with caution, in accordance with the "first, do no harm" doctrine.

8.2.6.4 Thermal Remediation

Thermal remediation utilizes heat to either mobilize or destroy the contaminants. The process by which contaminants are heated just enough to mobilize them and facilitate their removal from the soils is known as *thermal desorption*. The volatilized contaminants are then either collected or thermally destroyed. A thermal desorption system therefore has two major components: the desorber itself and the off-gas treatment system. The amount of heat needed to mobilize a contaminant depends on its chemical properties, with VOCs being the easiest compounds to treat (as usual).

In situ vitrification is the process by which the contaminated media is solidified by heating it to extremely high temperatures, in some cases above 1000°C. The result is the conversion of the porous, permeable soils or sediments into a solid mass of glass (as the name implies). The contaminants are either destroyed by the process or sealed within the vitreous mass, rendering them unable to migrate and therefore harmless. It can be implemented uniformly across a wide range of contaminants, although there must be some accommodation made for off-gases from the vitrification process (especially for PCBs, which can be converted into dioxins in an imperfect thermal destruction process).

8.2.6.5 In Situ Solidification

In situ solidification is similar in concept to *in situ* vitrification, since the process creates a solid mass in the subsurface, rendering the contaminants inert and harmless. Rather than utilizing heat, this process utilizes agents such as concrete mix to convert the soils into a solid mass.

8.2.6.6 In Situ *Chemical and Biological Treatment*

In situ chemical oxidation/reduction involves the injection of reagents into the contaminated area to induce oxidation or reduction, depending on the injected chemical. It is often coupled with *in situ* bioremediation, which can also be used to create an oxidizing or reducing environment as needed (see Section 8.2.6.7). The goal is to transform the contaminants into harmless compounds. The chemical reactions stimulate biological activity in which microorganisms metabolize (literally "eat") the contaminants, thus enhancing the transformation process. Table 8.1 summarizes the process by which contaminants are destroyed *in situ*.

The simplest example of biochemical treatment of contaminants involves metals and metal compounds. Since metals are elemental and cannot be transformed *in situ*, the goal of *in situ* metals remediation is to create a phase change through precipitation, by which the metal is changed into a solid, insoluble form and thus rendered immobile and harmless. Metals may be remediated by manipulating pH and creating either oxidation or reduction conditions to precipitate the metal contaminants. Chemical oxidation changes the oxidation state of the metal atom through the loss of electrons, whereas chemical reduction changes the oxidation state of metals by adding electrons. Typical oxidizing agents for metal contamination include potassium permanganate, hydrogen peroxide, hypochlorite, and chlorine gas. Typical reduction reagents include alkali metals, such as sodium and potassium, sulfur dioxide, sulfite salts, and ferrous sulfate.

Organic compounds, too, can be destroyed *in situ* by oxidative or reductive means. The endpoint of chemical oxidation is to reduce the contaminant to CO_2 and water. Chemical reduction is used to dehalogenate halogenated VOCs, thus creating harmless compounds. Biological oxidation and reduction have the same remediation endpoints.

To use a simple example, the following chemical equation shows how oxygen reacts with methane to form carbon dioxide and water:

$$CH_4 + 2O_2 \rightarrow CO_2 + 2H_2O$$

In this equation, the two oxygen atoms in the oxygen molecule "donate" electrons to the carbon atom and the hydrogen atoms in the methane molecule, enabling the formation of two new compounds. In the process, the oxygen atoms have "accepted" electrons from the methane atom. The processes of

TABLE 8.1

Four Options for *In Situ* Biochemical Remediation

	Bioremediation	Chemical Remediation
Oxidation	Bioremediation by oxidation	Chemical oxidation
Reduction	Bioremediation by reduction	Chemical reduction

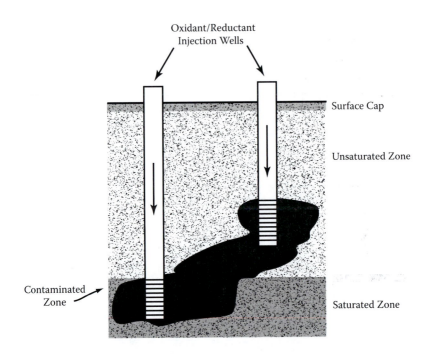

FIGURE 8.6
A schematic diagram of *in situ* chemical oxidation (ISCO).

electron donation and acceptance are complementary—one cannot exist without the other.

Aerobic microorganisms need oxygen to metabolize organic matter. Adding oxygen will stimulate the microorganism's ability to metabolize the contaminant by oxidation. Alternatively, anaerobic microorganisms need hydrogen to metabolize the contaminant by reduction. Oxidation and reduction, however, cannot take place simultaneously in the same area, due to the differing exigent chemistries involved.

Figure 8.6 shows the *in situ* chemical oxidation (ISCO) process. In this diagram, compounds containing electron donors or acceptors are injected into either the saturated or unsaturated zones. The most commonly used oxidation agents are ozone (O_3), hypochlorites, chlorine, chlorine dioxide, Fenton's reagent (hydrogen peroxide [H_2O_2] and iron), or potassium permanganate ($KMnO_4$). As suggested by their formulas, these active chemicals tend to have abundant oxygen in their molecular structure, enabling them to generate large quantities of free oxygen. The oxygen (or, in the case of Fenton's regeant, the radical created by the peroxide) then reacts with the contaminant, generating free radicals that energetically destroy contaminants. The result of the chemical oxidation process is the formation of harmless products such as CO_2, chlorine-containing compounds such as hydrochloric acid (HCl), and water.

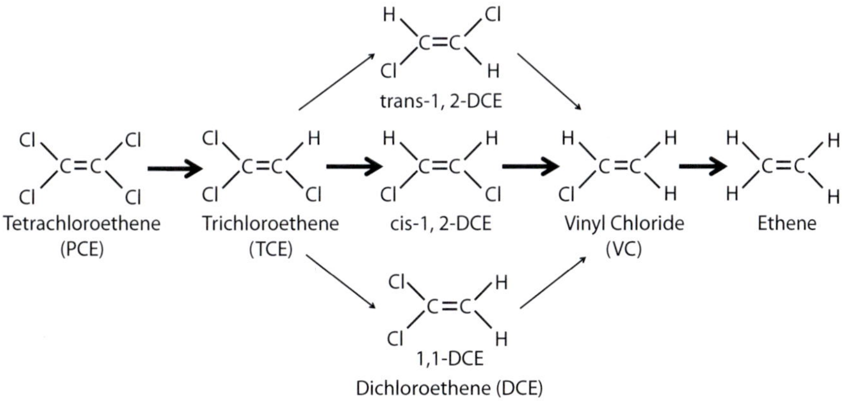

FIGURE 8.7
The reductive dechlorination process for perchloroethylene. (Courtesy of GZA GeoEnvironmental, Inc.)

The converse process invokes reductive conditions to remove halogens from the molecular structure. *Reductive dechlorination*, for instance, converts toxic chlorinated solvents, such as PCE and TCE, into harmless compounds, such as ethylene.

Figure 8.7 shows the reductive dechlorination process for PCE. In essence, this process reverses the method in which the PCE was formed, which was by taking ethylene, a two-carbon alkene structure and substituting chlorine atoms for hydrogen atoms. Reversing the process requires replacing a chlorine atom with a hydrogen atom, thus converting the PCE into TCE. Repeating this process four times will eventually yield harmless ethylene. Reductive dechlorination is aided by the presence of an electron donor substrate, which is a food source for microorganisms (an electron acceptor is oxygen or another substance that microbes use in digesting the electron donor substrate). This food source could be lactate, acetate, glucose, or some other simple organic compound that microorganisms can metabolize. The halogens, in this case chlorine, go free to form HCl and other such reaction products.

Chemical reduction can be enhanced by the introduction of certain nontoxic metals, such as iron, into the contaminated zone. Iron is an electron donor that transfers electrons, thus enhancing the reductive process.

The situation in the subsurface is rarely ideal, due to the presence of geologic complexities and chemicals that can interfere with the process through their usage of the oxygen intended to remediate the contaminant. As a result, the active ISCO ingredient often must be added in quantities that far exceed the theoretical quantity needed (and the quantity of the contaminant to be remediated) to effect the remediation.

8.2.6.7 *In Situ Bioremediation*

In situ bioremediation involves stimulating the biological activity portion of the remediation by introducing microorganisms (i.e., bioaugmentation) into the contaminated area. These microorganisms may be indigenous to the area, thus enhancing the established population, or foreign to the area (even genetically engineered), with properties that are desirable to the remediation of the contaminant of concern. As with chemical injections, *in situ* bioremediation involves the injection of water containing nutrients and alternate electron donors or acceptors, depending on whether aerobic or anaerobic degradation is the goal (food for the microorganisms) to stimulate the organism's growth and metabolic activity (see Figure 8.8). An impermeable surface cap helps control the injection process.

Many organic contaminants, particularly chlorinated compounds, may be toxic to natural bacteria. In recent years, bioengineering has created bacteria that are used in the bioremediation of chlorinated compounds because of their ability to withstand exposure to chlorine. Also, a class of naturally occurring bacteria, known as *methanotrophs*, is naturally resistant to chlorine and is cultured in laboratories for use as agent for *in situ* bioremediation.

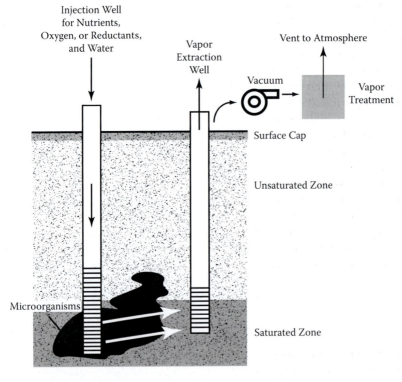

FIGURE 8.8
A schematic diagram of *in situ* bioremediation.

Bioremediation also can be accomplished *ex situ* by excavating contaminated soils and mixing microorganisms into the soil pile, usually as part of an aqueous slurry, while aerating it to increase the amount of oxygen available to the microorganisms. The mixture of contaminated soil and slurry is known as a *biopile*. *Ex situ* bioremediation is less effective in low temperature environments, which will slow the metabolism of the microorganisms.

Pilot- or bench-scale testing is important for bioremediation, especially for reductive dechlorination. Vinyl chloride, the penultimate step in the dechlorination process, is highly toxic to humans, with similar toxicity to PCE and TCE. It is also more stable than those compounds, so once vinyl chloride is created, it becomes difficult to complete the dechlorination process. Small-scale testing prior to implementing the full remediation would assess whether the proposed bioremediation technique will achieve full dechlorination or stop short at an unwanted endpoint.

8.2.6.8 Bioventing

Bioventing entails the bioremediation of contaminants by pumping air directly into the contaminanted portion of the unsaturated zone through injection wells (see Figure 8.9). Periodically, nutrients are added to the injected air, which enhances the natural biodegradation process. Vapor recovery wells screened in the unsaturated zone are used to recover vapors that may contain contaminants, although this aspect of the remediation is of secondary concern. As with many other *in situ* remediation methods, it is important to maintain an impermeable cap above the zone of remediation to avoid short-circuiting.

8.2.6.9 Permeable Reactive Barriers

A *permeable reactive barrier* is a "wall" constructed in the subsurface through which groundwater can flow (see Figure 8.10). This process, sometimes known as *funnel and gate*, requires a detailed knowledge of the geologic and hydrogeologic conditions in the area of the contaminant plume.

The reactive medium contains chemicals designed to react with the contaminants dissolved in the groundwater, creating harmless substances such as carbon dioxide and water. It can be a mix of physical, chemical, or biological processes. It may contain GAC to capture organic contaminants, certain metals that will react with the contaminants or act as catalysts by causing the dehalogenation of contaminants, or microorganisms and electron acceptors.

A reactive barrier typically is constructed by digging a trench deep enough and wide enough to intercept the entire contaminant plume. Once constructed, the trench is backfilled with the reactive material. Impermeable slurry walls are sometimes constructed along the flanks of the reactive

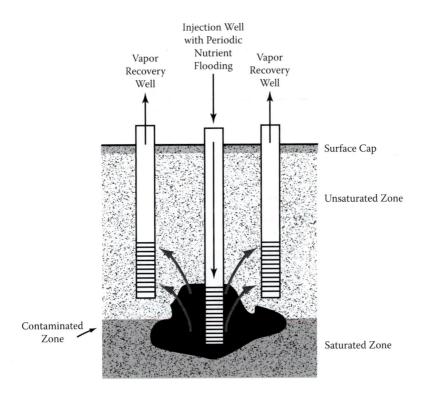

FIGURE 8.9
A schematic diagram of a bioventing system.

barrier, creating a channel, or a funnel, that directs the contaminant plume toward the reactive wall.

8.3 Landfill Closures

The closure of uncontrolled landfills presents special circumstances to the remediation engineer due to the sheer quantity and variety of wastes involved. Landfill closures typically involve three main elements to control contaminant migration from the landfill: capping, venting, and leachate collection.

Capping involves the installation of an impermeable cap across the landfill. The cap should be designed to account for settling of the underlying materials, and needs to be monitored periodically to ensure its integrity. HDPE liners (see Section 8.2.1) are often a component of a landfill cap.

Organic materials deposited in the landfill continue to degrade even once waste deposition ceases. Once the landfill is capped, oxygen can no longer penetrate into the landfill, resulting in the creation of an anoxic environment.

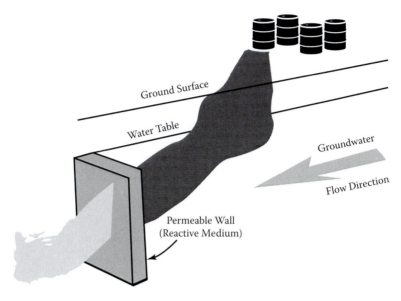

FIGURE 8.10
A schematic diagram of a permeable reactive barrier.

This results in anaerobic degradation of organic matter, which creates methane. Methane is explosive and therefore must be controlled to avoid it building up to dangerous concentrations within the landfill or nearby structure.

To control the buildup of methane within a landfill, a *methane venting system* is constructed. This system consists of a permeable gas collection layer, built into the surface cap, which is designed to collect methane that is migrating to the surface. It is constructed beneath an impermeable barrier to enhance gas collection. This layer is analogous to the permeable layer constructed beneath a building structure as part of a radon mitigation system (see Chapter 16). Vents are used to channel the methane to the surface. They may be situated in the gas collection layer, beneath this layer within the landfill, or both. These vents may be passive or active, in which case vacuum pumps attached to the vent pump the methane to a surface collection system. In some cases, the recovered methane can be used as fuel for a cogeneration energy system.

As the wastes within the landfill settle, the wastes tend to dry out, creating free liquids that, due to the impermeable cap, cannot vaporize to the atmosphere. Instead, they condense and travel downward in the landfill and, if left uncontrolled, will carry contaminants outside of the landfilled area. To control free liquids generated within landfills, all controlled landfills, hazardous or otherwise, are designed with *leachate collection systems*. These systems consist of underground piping that are designed to capture leachate and transport it to a wastewater treatment system.

8.4 Remediation and Consultants

Consultants play many roles in the remediation of soils and groundwater. Risk assessments are usually performed by consultants with backgrounds in toxicology, chemistry, and statistics. Remediation systems usually are designed by environmental engineers, although an experienced nonengineer could design most remediation systems.

Remediations that involve heavy machinery used for trenching, excavation, and so forth are usually performed by contractors, with consulting engineers observing the field activities to ensure that they follow the work plan. Consultants may also perform air monitoring during the remediation activities, especially excavation activities, to assess whether hazardous vapors are escaping the work area and potentially affecting workers and neighboring properties. This type of air monitoring, sometimes called *community air monitoring*, is often performed along the property boundary using either handheld monitoring devices or automatic data loggers.

Follow-up work, involving post-excavation sampling, groundwater monitoring, etc., is usually performed by consultants. As in most cases, the paperwork, remedial action work plans, permit applications, and remedial action reports (prepared at the conclusion of the remedial activities) are the province of environmental consultants.

References

Evanko, Cynthia R., and Dzombak, David A. October 1997. Remediation of Metals-Contaminated Soils and Groundwater. Technology Evaluation Report TE-97-01. Ground-Water Remediation Technologies Analysis Center.

Hutzler, Neil, Murphy, Blane E., and Gierke, John S. 1989. State of Technology Review: Soil Vapor Extraction Systems. EPA/600/2-89/024. EPA Risk Reduction Engineering Laboratory.

Interstate Technology Regulatory Council (ITRC). December 2009. Evaluating LNAPL Remedial Technologies for Achieving Project Goals.

Koenigsberg, Stephen S., ed. 2004. Cost-Effective Groundwater Remediation: Selected Battelle Conference Papers 2003–2004. Regenesis.

National Research Council. 1994. *Alternatives for Ground Water Cleanup.* Washington, DC: National Academy Press.

New York State Department of Environmental Conservation. May 2010. DER-10: Technical Guidance for Site Investigation and Remediation.

Nyer, Evan K., Palmer, Peter L., Carman, Eric P., Boettcher, Gary, Bedessem, James M., Lenzo, Frank, Crossman, Tom L., Rorech, Gregory J., and Kidd, Donald F. 2001. *In Situ Treatment Technology,* 2nd ed. Boca Raton, FL: CRC Press.

Testa, Stephen M., and Winegardner, Duane L. 2000. *Restoration of Contaminated Aquifers*, 2nd ed. Boca Raton, FL: CRC Press.

U.S. Environmental Protection Agency. September 1998. Technical Protocol for Evaluating Natural Attenuation of Chlorinated Solvents in Ground Water. EPA/600/R-98/128. Office of Research and Development.

U.S. Environmental Protection Agency CLU-IN. Technologies: Remediation, http://clu-in.org/remediation.

Watts, Richard J. 1997. *Hazardous Wastes: Sources, Pathways, Receptors*. New York: John Wiley & Sons.

Wiedemeier, Todd, Kampbell, Don H., Ferrey, Mark, and Estuesta, Paul. 2001. Technical Protocol for Implementing Intrinsic Remediation with Long-Term Monitoring for Natural Attenuation of Fuel Contamination Dissolved in Groundwater. Air Force Center for Environmental Excellence.

Section III

Land Usage

9

Wetlands, Surface Waters, and Endangered Species

The ultimate objective of environmental investigations and remediations is to protect human health and the environment. Chapter 7 discusses generic remediation standards and human risk assessment methods designed to quantify remediation goals that will be protective of human health. In this chapter, we discuss the development of remediation goals designed to be protective of the environment. Before discussing *ecological risk assessment*, it is necessary to discuss the myriad of considerations in understanding the nonhuman environment, and the parameters and tools used in assessing its condition. The chapter concludes with various mitigation methods that are typically employed to restore damaged portions of the nonhuman environment.

9.1 Ecosystems

An *ecosystem* can be defined in many ways. For our purposes, we define an ecosystem as the interactions between microbial, plant, or animal life as well as the nonliving (abiotic), physical components of the environment with which they interact, such as air, soil, sunlight, and, most especially, water. A healthy ecosystem is recognizable by its diversity and abundance of animal and plant life. Lack of animal and plant diversity, abnormally small or struggling populations of indigenous species, or the presence of invasive species that are crowding out the native species are all signs of an unhealthy ecosystem.

Ecosystems can be damaged by changing one of the following factors:

- The amount of water available to the system (such as through water diversion, river channelization, or natural means, such as flood or drought)
- The configuration of the system, such as through the construction of dams, infilling of surface waters, and so forth
- The system chemistry through the introduction of chemicals into the system or the changing of basic water chemistry parameters

- The biology of the system through the introduction of nonnative species or destruction of native species

Ecological investigations are conducted for several reasons. They may be conducted as part of a contamination investigation, in which the investigator assesses whether a release of hazardous wastes into the environment has impacted an *ecological receptor*, defined as a plant or an animal other than a human. This investigation may occur as part of an ecological risk assessment, as a required step under the National Environmental Protection Act (NEPA) and state equivalents during the planning stage of a construction project (see Chapter 10), or to determine the health of the ecosystem unrelated to a spill event or planned action. The identification and delineation of wetlands, an ecological receptor, is a required step under the Section 404 of the Clean Water Act.

Ecological receptors may be classified in one of the following categories, each of which is discussed in this chapter:

- Surface water
- Sediments
- Wetlands
- Flora and fauna

9.2 Surface Water and Sediment Investigation

9.2.1 Pollution Sources

Sources of surface water and sediment pollution generally fall into two categories: point source and nonpoint source. *Point sources,* as the name implies, originate from a particular point or place, for example, an accidental release from an oil storage facility or a wastewater pipe outfall (see Figure 9.1). *Nonpoint sources* (NPS) are caused by rainfall or snowmelt moving over and through the ground. While moving over and through the ground, they pick up contaminants and deposit them in nearby water bodies. Typical contaminants from nonpoint sources include fertilizer and pesticide runoff from a farm; chemicals settling out of the air; chemicals transported downstream from a variety of sources or no discernable source; and diffuse anthropogenic pollution (DAP), such as deposits from motor vehicle emissions and combustion of fossil fuels for heating purposes. Nonpoint sources, in general, may constitute a larger, perhaps regional problem that can be hard to diagnose and difficult to mitigate.

FIGURE 9.1
A storm water outfall pipe. (Courtesy of GZA GeoEnvironmental, Inc.)

9.2.2 Surface Water Indicator Parameters

There are various inorganic parameters that are commonly used to measure surface water quality. Among these parameters are the following:

Biochemical oxygen demand (BOD)—BOD measures the amount of dissolved oxygen needed by aerobic organisms to stabilize the organic material present in a body of water. Soluble organic matter introduced into a surface water body will increase its BOD, resulting in oxygen depletion in the water and the suffocation of aquatic animals. Animal waste is the most common organic matter to increase BOD. Animal waste includes human waste originating from a septic system or from a malfunctioning municipal sanitary sewer system, and waste from livestock (cattle, pigs, sheep, etc.) in rural areas. It is measured in milligrams per liter (mg/l) or the equivalent parts per million (ppm).

Chemical oxygen demand (COD)—COD measures the oxygen demanded by chemicals present in the water body. These chemicals, in some cases domestic and industrial wastes introduced into the water body, compete with aquatic animal life for available oxygen. High COD, like high BOD, can result in the suffocation of aquatic animals. COD also is measured in milligrams per liter and parts per million.

Dissolved oxygen—Dissolved oxygen (DO) is the amount of oxygen available in water to aquatic animals. DO values of less than 5 ppm typically are considered detrimental to fish.

Total organic carbon (TOC)—TOC measures the amount of organic carbon in the aquatic system. It does not include inorganic carbon, usually in the form of carbon dioxide and carbonic acid salts. TOC indicates the amount of decayed organic matter in the water body, which is indicative of the health of the organisms in the water body.

Turbidity—Turbidity measures the opacity of the water. It is related to the presence of suspended solids in the water and is recorded as nephelometric turbidity units (NTUs). High turbidity could block sunlight for aquatic plants or can silt up the water body, reducing its capacity.

Total solids—Total solids is the sum of suspended solids (particulates in the water) and dissolved solids. High total solids in a surface water body is usually caused by high suspended solids, which is indicative of soil erosion or other infiltration of materials.

pH—Also known as corrosivity, pH is generally stable in a healthy ecosystem, and most aquatic ecosystems infringing on the normal human environment range from a pH of 5 to 9. Abnormally low or high pH not only affects organisms directly, but it also affects the solubility of various metals (see below), which can also affect their chemical state, mobility, biological/toxicological effects, and the entire ecosystem.

Salinity—Salinity is the concentration of chlorides [Cl^-] in the water that cause it to have a salty taste. Water with a salinity reading of less than 0.5% is considered fresh water. Excessive salinity can impact fresh water flora and fauna, and can increase the solubility of certain undesirable compounds, such as metals that may impact surface water quality. Conversely, a decrease in salinity can cause these same metals to precipitate out of water and fall to the bottom of the water body, thereby impacting the sediments at the bottom of the water body.

Conductivity—Conductivity (not to be confused with hydraulic conductivity, see Chapter 4) is the ability of the water body to conduct electricity. It is a function of the concentration of ions and particulates within the water. It is related to salinity and is usually measured in micromhos per centimeter (μmhos/cm).

Hardness—Also known as alkalinity, hardness is measured by the concentration of calcium carbonate ($CaCO_3$) in the water. Measuring "hardness" is important, especially if metals analyses are to be performed, since metals are more toxic in "soft" water (low calcium concentration) than "hard" water (high calcium concentration).

Various metals, including iron and manganese—The presence of various metals can raise the COD of a water body.

Heavy metals, such as lead, chromium, copper, and zinc—Heavy metals, present at natural low level concentrations in most aquatic environments, typically reach toxic levels from the release of industrial wastes. They have the ability to kill flora and fauna, and inhibit living flora and fauna from reproducing.

Sulfides (S^{2-})—Sulfide, an anion of sulfur, is highly basic and can raise the pH of a water body. Sulfides are formed in oxygen-deficient environments by anaerobic microbial respiration using sulfates (SO_4^{2-}) as an electron receptor instead of oxygen. Sulfides are often seen in natural anaerobic environments such as wetlands (see Section 9.5) or anoxic water bodies.

Phosphorus (P) and nitrogen (such as nitrates [NO_3^-] and nitrites [NO_2^-])—Phosphorus and nitrogen are nutrients necessary for all plant and animal life. High levels of phosphorus and nitrogen accelerate an ecosystem's productivity, resulting in a sharp growth in aquatic plant life, such as the algal bloom shown in Figure 9.2, or the spread of invasive and nuisance aquatic plants. This is known as *cultural eutrophication*, that is, an increase in plant life (eutrophication) due to human (cultural) activities. The blooms can block sunlight, killing off aquatic plants. Certain algal species can also produce toxins that can affect other plant and animal life, including humans (e.g., red-tides algae, cyanobacteria). Excessive phosphorus and nitrogen often originate from fertilizers that are transported into a water body by storm water runoff. Once their introduction is brought under control, the algae die and decompose, resulting in an oxygen-deficient environment that can kill aquatic animals.

As this discussion suggests, many of these measurements affect other measurements. Therefore, in most cases it is important to collect many kinds of data to gain an understanding of the status of a water body.

9.2.3 Surface Water Sampling

Surface water and sediments should be sampled if discharges of contaminants have occurred or are expected to have occurred. Evidence of stressed vegetation, sheens on the water or sediments, seeps, or discolored soils or sediment are indicative of present or past spills. Evidence of fish kills, or other degradation of flora and fauna populations may also trigger the acquisition of site-specific data to support the investigation.

In the case of suspected point source pollution, samples should be collected as close to the suspected point of deposition, which is usually the point where the water is moving the slowest. Bracketing samples should be collected upstream and downstream of the suspected point source. The

FIGURE 9.2
Algal bloom in a river in Sichuan, China. (Courtesy of Felix Andrews.)

upstream sample provides quality control by determining background contaminant concentrations.

Let's go back to the fictitious manufacturing facility and add in a wastewater pipe (see Figure 9.3). Once wastewater discharges from a pipe and encounters the atmosphere, there will be a loss of some volatile constituents. Most contaminants, however, will mix with the receiving stream water and be carried downstream. Any nonaqueous phase constituents in the wastewater may dissolve, but most chemicals will be in the dissolved phase. These chemicals will move downstream, although chemicals with high soil–water coefficients or high organic carbon coefficients may adhere to the underlying sediment in the stream. In addition, under some hydraulic conditions chemicals may work their way into the underlying soils beneath the stream.

An appropriate sampling scheme for the surface water and sediment in the brook would include sampling locations upstream, downstream, and at the location of the point discharge. The surface water sample assesses the presence of dissolved chemicals, and the sediment sample tests for chemicals

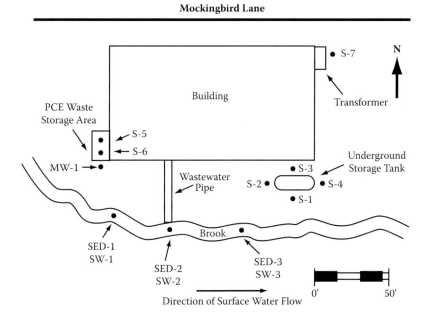

FIGURE 9.3
Surface water and sediment sampling to test for point source pollution emanating from a wastewater pipe at a manufacturing facility.

that have sorbed onto soil particles. The water sample should be analyzed for the contaminants suspected to be present in the point source release.

To assess surface water quality or the effects of nonpoint pollution sources on the water body, a more thorough sampling of the water body is warranted. Since there are wide variations in the types and sizes of surface water bodies, there is no one-size-fits-all sampling procedure. However, there are certain universal principles that will aid in the proper design of a sampling and analysis plan (SAP).

As a general rule, the quantity of water samples to be collected depends on the size of the water body and the expected chemical and biological variations within the water body. For instance, one sampling point is usually sufficient for narrow or shallow streams that are suspected to have homogeneous conditions along the stream. For deeper and wider streams, more sampling locations from various "transects" (cross-sections of the stream) are necessary because physical conditions can vary horizontally and vertically, which can affect the chemistry of the water.

The amount of interaction with the atmosphere varies with depth in deeper water bodies. Water samples from middle or lower depths of the water body will avoid sampling water that has interacted with the atmosphere and therefore may not be representative of the entire water body. On the other hand, there may not be such a thing as a representative sample in a water body with vertical circulation patterns, or influences from discharges, tributaries, and so forth. Depending on the goals of the investigation, samples from multiple depths may need to be collected. Temperature readings at varied depths in the water body can indicate whether the water chemistry is sufficiently different by depth to warrant multiple samples.

When collecting multiple samples from a water body, consideration should be given to the effect of one sample on the subsequent samples. For instance, the act of collecting a sample in a river could agitate the water that is heading downstream and thus affect subsequent downstream sampling. Therefore, sampling should proceed from downstream to upstream locations. Similarly, sampling should proceed from the top of a water body to the bottom of the water body.

Collecting samples at different times often is needed to characterize the water body. The effects of seasonal weather (cold vs. hot, wet vs. dry) and related biological variations; high flow and low flow stages for rivers and streams; and tidal variations for applicable water bodies can lead to chemical variations that cannot be measured or understood in one sampling event.

The quality and quantity of water flow is often an important parameter in assessing an ecosystem that is tied to a river or stream. High flow rates, or turbulent flow, can affect chemical measurements, especially for volatile organic compounds (VOCs). In addition, the rate of flow can affect sample collection, since it is often impossible to collect a water sample in the middle of a fast-flowing river. Flow is recorded as volume per unit time and is expressed in units such as cubic feet per second (CFS), gallons per minute (GPM), or millions of gallons per day (MGD).

9.2.4 Sediment Sampling

Contaminated sediments can be a repository for contaminants, which can be remobilized back into the overlying water body, and a conduit for pollution into the subsurface and pathway for pollutants into the underlying aquifer (see Chapter 4). Last, sediments can provide a pathway by which contaminants enter the food chain through flora and bottom-feeding fauna, which can in turn be a source of food for fish, small mammals, and birds. Once consumed, the contaminants can work their way up the food chain by the process of *bioaccumulation.*

Because of the dangers of bioaccumulation to aquatic animals, sediment sampling and analysis should also focus upon those chemicals that are persistent in the aquatic environment, have high bioaccumulation potential, are toxic or teratogenic (causing birth defects) to aquatic organisms, and have a high frequency of detection. Investigations for bioaccumulative chemicals in sediments can be enhanced by collecting fish and other aquatic animals, and performing tissue analysis for the presence of such chemicals.

In some ways, the goals of sediment sampling are similar to soil and groundwater sampling, which are the identification and delineation of contamination (see Chapters 6 and 7). In the case of sediment contamination, it is also important to establish background conditions, since, without that knowledge, it may be difficult or impossible to assess whether the detected contamination originates from the point source or from background (i.e., nonpoint source) conditions. At a minimum, samples should be collected from the top several inches of sediment, possibly supplemented by deeper samples. Collecting measurements for TOC, pH, and particle grain size aids in the understanding of contaminant migration through the sediments.

9.3 Assessing Impacts on Fauna

Impacts of contaminants on fauna can be modeled in many ways, as described in the next section. Quantitative data can be obtained by conducting a tissue residue study, which entails collecting samples of the animal in question and submitting a sample of its tissue for laboratory analysis. Conducting tissue residue studies on the various animals within a food chain will help the investigator estimate the bioaccumulative effects of contaminant uptake within the ecosystem.

9.4 Ecological Risk Assessment

As with soil and groundwater sampling and analysis, generic standards usually are available by which to compare analytical results to establish whether the concentration of contaminants are "acceptable" in that they do not pose a threat to the environment. However, because of the myriad of variations in environmental settings, these standards can often be misleading. Therefore, where ecological impacts are suspect, an ecological risk assessment is performed.

An ecological risk assessment is analogous to a human health risk assessment in that its goal is to establish levels of contaminants that will not have a detrimental effect on the subject fauna and flora. The contaminants of concern are generally the same for both types of risk assessment, although the levels of concern signified by certain contaminants differ.

The environmental risk assessment can be thought of containing three phases (NJDEP, 2011):

Phase I—Problem Formulation

Phase II—Analysis

Phase III—Risk Characterization

9.4.1 Problem Formulation

The goal of Phase I—Problem Formulation is to develop an *ecological conceptual site model* (ECSM). The ECSM is similar to the conceptual site model described in Chapter 6. Possible source-to-pathway-to-receptor scenarios are developed for the various contaminants of concern in the study area.

Exposure routes to fauna include three of the same exposure routes as with humans: ingestion, direct contact, and inhalation (injection is not considered an exposure route for nonhuman animals). However, there is an additional exposure route for nonhuman animals: bioaccumulation, which represents an indirect exposure route through the consumption of contaminated prey.

Preliminary data is collected to investigate the potential exposures to receptors under the various scenarios. The goal of the data collection is to establish *measurement endpoints,* that is, the results of sampling and analyses that are used to estimate the exposure, effects, and characteristics of the ecosystem. These measurement endpoints are as varied as the ecosystem itself and are difficult to generalize. They include measurements of the parameters discussed earlier in this chapter for surface water, sediment, and wildlife sampling, as well as parameters specific to a given scenario.

9.4.2 Analysis

The analysis phase involves technical evaluation of the data collected in Phase I. It includes estimating potential contaminant exposures to receptors and their potential effects on the receptors. All four of the aforementioned exposure routes must be considered in the analysis phase.

9.4.3 Risk Characterization

The risk characterization phase of the ecological risk assessment is similar to the Risk Characterization phase of the human health risk assessment. Its goal is to estimate the likelihood of ecological risks by examining each exposure scenario and evaluating the potential effects on flora and fauna for each assessment endpoint. This phase includes a *toxicity assessment* (see Chapter 7), which involves:

- Research into the latest toxicological studies to ensure that the most recent research involving the chemicals of concern is employed in the risk assessment process

- Performing biological surveys to assess actual area conditions
- Performing toxicity tests on actual media

A key component of ecological risk characterization is *food chain modeling*, which is used to predict how contaminants work their way up the food chain. The pathways considered in food chain modeling include bioaccumulation as well as direct consumption of contaminated water, soil, or sediment. The total potential exposure for the animal in question to the contaminant, known as the *potential dose*, will be the sum of these three pathways.

Estimating the contribution of bioaccumulation to the total potential exposure is done by conducting a *bioaccumulation (uptake) study*. This includes a tissue residue study, as described in Section 9.3. Alternatively, the investigator can use generic *sediment/soils-to-biota bioaccumulation factors* (BSAFs) and *bioaccumulative factors* (BAFs) that are published by the Army Corps of Engineers (http://el.erdc.usace.army.mil/basnew), the U.S. Environmental Protection Agency (www.epa.gov/med/Prod_Pubs/bsaf), and other sources.

Once each assessment endpoint has been calculated, they are compared to *toxicity reference values* (TRVs), which are similar to reference doses and reference concentrations described in Chapter 7. If the assessment endpoint is greater than the TRV, then an ecological hazard exists. In such cases, the contamination must be remediated or, in the case of a proposed action, must be reconsidered in light of the potential ecological hazard.

Because biological systems are variable, the uncertainty surrounding estimates of ecological risk may be much greater than those associated with a human health risk assessment. The investigator should account for this variability when evaluating the results of the ecological risk assessment.

9.5 Wetlands Identification and Delineation

The U.S. Environmental Protection Agency (USEPA) identifies the following six categories of special aquatic sites in its Clean Water Act Section 404 guidelines:

- Sanctuaries and refuges
- Wetlands
- Mudflats
- Vegetated shallows
- Coral reefs
- Riffle and pool complexes

Of these six categories, wetlands merits special attention due to its prevalence throughout the United States and its impact on real estate and development.

9.5.1 Definition of Wetlands

Section 404 of the Clean Water Act defines *wetlands* as

> those areas that are inundated or saturated by surface or ground water at a frequency and duration sufficient to support, and that under normal circumstances do support, a prevalence of vegetation typically adapted for life in saturated soil conditions. Wetlands generally include swamps, marshes, bogs, and similar areas.

This definition contains phrases that need to be clarified.

Under normal circumstances—This does not include areas disturbed by human influences and conditions at times of extreme droughts or similar events.

A prevalence of vegetation—Vegetation associated with wetlands must be present but not exclusively present. Conversely, the mere presence of vegetation associated with wetlands does not mean wetlands are present, especially if it is not the prevalent form of vegetation in the area. This imprecise term leaves much to the professional judgment on the part of the regulators and the environmental consultant.

Typically adapted—Vegetation that is normally suited to the conditions in the wetlands area.

There is no minimum size that defines a wetland. The following sections define the conditions needed in wetlands.

9.5.2 Wetlands Hydrology

Wetlands need not be wet all of the time and may not often even have visibly standing water. The frequency and duration of wetness depends upon many factors. However, one thing all wetlands have in common is the presence of sufficient water for the surface soils to stay sufficiently wet to create anaerobic conditions in the soil for a significant portion of the growing season. Either surface water, or groundwater at an elevation no more than 14 inches (4.7 cm) below the soil surface, should be present for at least several weeks of the growing season.

Most surface soils have aerobic conditions due to the ready flow of air through the unsaturated pore spaces of the soil. However, when soils become saturated with water due to rainfall and flooding, the water impedes the flow of gases, including oxygen, through the soil. Microbial activities also act to deplete the soils of oxygen, creating *hydric* conditions. *Hydric soil* is "a soil that formed under conditions of saturation, flooding, or ponding long enough during the growing season to develop anaerobic conditions in the upper part" (*Federal Register*, July 13, 1994). The National Technical Committee for Hydric Soils (NTCHS) has the responsibility to specifically define and develop or accept criteria for hydric soils.

Hydric soils have been saturated often and over many years, allowing them to develop unique properties that can be recognized in the field. Because of the lack of oxygen in the soils, reducing rather than oxidizing conditions prevail. Reducing conditions result in the creation of iron and manganese oxides, which give hydric soils their distinctive coloring. *Gleyed soils*, a category of hydric soils, have a distinctive grey coloring, sometimes greenish or bluish grey. Also, hydric soils often have *mottles* (also called redoximorphic accretions and depletions), which are spots of contrasting color.

Hydric soils support hydrophytic vegetation. If drained, by either human activity, such as the construction of dams or levees, agricultural tile drainage, or some other impediment to natural surface water flow, or by natural conditions, such as prolonged drought, they may no longer qualify as hydric soils.

Water is introduced into an area by precipitation, natural drainage patterns, periodic flooding, and high water tables. The ability of wetlands to stay wet depends upon the amount of water introduced into the area and the soil's ability to retain that water. Soil's ability to retain water depends upon the types of soils, their porosity, and their permeability. Beach sand, for instance, may become saturated constantly by wave action. However, as anybody who has been to the beach knows, the water quickly drains from the sand once a wave recedes and becomes dry in a matter of minutes if not wetted again. Clays and hardpan soils are common to wetland environments because of their ability to retain water. However, sandy wetlands are not uncommon, but depend upon high groundwater levels present for a significant portion of the growing season.

9.5.3 Wetlands Vegetation

Wetlands vegetation is defined by the ability of the individual species of plant to actively grow and reproduce in areas with hydric soils (i.e., an anaerobic environment). There are various wetland plant communities, which are species that are naturally associated with each other. Vegetation associated with wetlands is known as *hydrophytic* vegetation, that is, vegetation typically adapted for life in saturated soil conditions. The presence of scattered individual plants does not make a community and cannot be used for wetlands classification.

9.5.4 Wetlands Indicators

For a wetlands to be present, positive *wetlands indicators* of hydrology, types of soils, and types of vegetation must be present. Sole reliance on the presence of wetlands vegetation can be misleading, as many plant species commonly associated with wetlands can grow successfully on nonwetlands as well. Furthermore, hydrophytic vegetation and hydric soil conditions could persist for decades after the alteration of the area hydrology rendered the area a nonwetlands.

9.5.5 Classifying Wetlands

Wetlands are classified by system, which describes the overall environmental setting, and subsystem, which describes the frequency of inundation. There are four wetlands systems: marine (oceanic), estuarine (associated with an estuary), riverine (associated with a river), and palustrine (not directly associated with a surface water body). These systems and their subsystems are described next.

Marine systems are oceanic wetlands. There are two types of marine wetlands. Marine "subtidal" refers to wetlands in which the substrate, that is, the portion of the soil that supports life, is constantly submerged. Marine "subtidal" refers to wetlands in which the substrate is exposed by tides.

Estuarine systems are deepwater tidal habitats and adjacent tidal wetlands that are connected to the ocean or other large saltwater body, although this connection is obstructed to varying degrees by land (see Figure 9.4). The water in estuarine wetlands is at least occasionally diluted by freshwater runoff from the land. Similar to marine systems, estuarine "intertidal" lands are constantly submerged, and estuarine "intertidal" lands are exposed by tides.

Riverine systems include all wetlands and deepwater habitats contained within a channel, such as a river, excepting wetlands dominated by trees, shrubs, and so forth, and habitats with brackish water. There are four riverine subsystems. Riverine "tidal" lands are periodically exposed by tides. There are two riverine perennial subsystems, distinguished by a lack of tidal influence and water flow throughout the year. In riverine perennial lower subsystems, the flow gradient is low and the water velocity is slow. In a riverine upper perennial subsystem, flow gradient is high and water fast. In riverine intermittent wetlands, the water body flows only part of the year.

Lacustrine systems are wetlands and deepwater habitats that are situated in a topographic depression (such as a lake) or a dammed river channel that is larger than 20 acres (80,000 m²) in area. The water

FIGURE 9.4
A typical marshy area. (Courtesy of GZA GeoEnvironmental, Inc.)

body lacks trees, shrubs, persistent "emergent" plants (i.e., plants that are rooted below the water but extend above the water surface), and emergent mosses or lichens with greater than 30% areal coverage. Lacustrine systems include permanently flooded lakes, reservoirs, intermittent lakes, and tidal lakes with fresh water.

Deepwater habitats within the lacustrine system are known as *limnetic* lacustrine systems. Wetlands habitats within the lacustrine system are known as littoral lacustrine systems. They extend from the shoreward boundary of the system to a depth of 2 m (6.6 feet) below low water or to the maximum extent of nonpersistent emergent plants.

Palustrine systems include all nontidal wetlands that are dominated by trees, shrubs, persistent emergent plants, and emergent mosses or lichens, as well as all such wetlands that occur in freshwater tidal areas. Palustrine wetlands range from permanently saturated or flooded land (as in marshes, swamps, and lake shores) to land that is wet only seasonally (as in vernal pools).

9.5.6 Available Information on Wetlands

The United States Army Corps of Engineers (USACE) Wetlands Delineation Manual describes the procedures for the identification and delineation of jurisdictional wetlands, that is, those wetlands that are regulated under Section 404 of the Clean Water Act. Before attempting to delineate wetlands on a particular property, however, it is important, as it is for so many environmental investigations, to do your homework first.

There are ample resources regarding wetlands available for free and online. Most important is the *National Wetlands Inventory*, which was developed as a response to the Emergency Wetlands Resources Act of 1986 and subsequent amendments. The National Wetlands Inventory has produced maps covering almost 90% of the conterminous United States, portions of Alaska, and all of Hawaii and the U.S. Territories. Thousands of these maps have been digitized and are available online at www.fws.gov/wetlands. States, county, and local agencies, such as planning and engineering departments, also have data on wetlands locations within their jurisdictions. Most states now have for download and general usage geographic information system (GIS) layers showing wetland and hydrologic boundaries, and soil classifications. Other available resources include:

- United States Geological Survey (USGS) topographic maps, which show the general locations of wetlands.
- The Natural Resources Conservation Service (formerly the Soil Conservation Service), a division of the United States Department of Agriculture, which has published soil surveys by county for most of the United States (http://soils.usda.gov/survey). The location of soil types and their general characteristics are defined in the soil surveys.
- Abundant ecological information can be obtained from environmental impact statements (see Chapter 10) and similar publications.
- Aerial photographs from public and private sources provide general information about the presence of wetlands. This is particularly true of infrared aerial photography in which red is indicative of live vegetation and the tone of red can be a guide to the density and health of the vegetation and how vigorously it is growing. Dead vegetation usually appears as various shades of tan or green on infrared photographs.
- The U.S. Fish and Wildlife Service (USFWS) has a comprehensive list of the thousands of species of hydrophytic plants, with a cross-reference to the geographic area where they appear. It can be accessed at http://library.fws.gov/Pubs9/wetlands_plantlist96.pdf. Knowledge of wetlands plants in the local area is critical to an accurate mapping of wetlands.
- Studies performed for public and private entities in support of a proposed development, such as the wetlands delineation map shown in Figure 9.5.

As important as these resources are for identifying wetlands, they should never be considered as authoritative regarding an area or certainly a specific property, given the gross scale by which these data sources are provided and

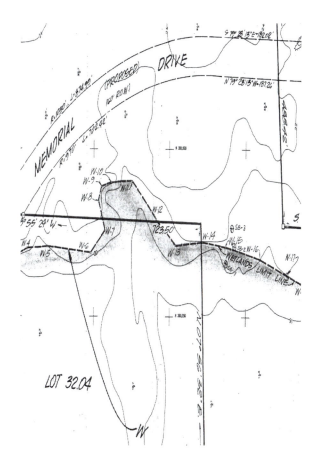

FIGURE 9.5
A portion of a wetlands map, showing the location of wetlands and their soil classification.

their frequent inaccuracies. When it comes down to identifying and delineating wetlands in an area or on a specific property, there is no substitute for the collection of field data.

9.5.7 Field Mapping of Wetlands

Once the publicly available data are accessed and reviewed, the investigator,[*] armed with a compass, a measuring device, and a shovel/auger, is ready to go into the field (see Figure 9.6).

The investigator will undertake several traverses of the subject area looking for evidence of inundation and soil saturation, and identifying plant communities and dominant species. When soils are submerged, the investigator will note the depth of submergence and the plant communities present.

---

[*] Wetlands investigators must be licensed to operate in certain states.

FIGURE 9.6
A wetlands investigator in the field. (Courtesy of GZA GeoEnvironmental, Inc.)

When the soils are not submerged, the investigator may determine the depth that water is present by digging a hole, known as a "soil pit," to a depth of 24 inches (60 cm). If water drains into the hole after a certain period of time has elapsed, it is an indication of shallow saturated soils, suggesting that the ground may be submerged during periods of heavy precipitation or flooding. Knowledge of hydrologic cycles in the area, including tidal influences, as applicable, assists the investigator in determining whether the field observations suggest the presence of wetlands. The investigator should also note whether the soils have characteristics of hydric soils.

Some clues regarding the persistence of a high water table, or the existence of conditions suggesting the presence of wetlands, include but are not limited to the following:

- The presence of living or dead aquatic fauna
- Watermarks on trees
- Morphological adaptations in vegetation

- The presence of an algal mat or crust
- Drift lines, which is an accumulation of debris from a water flow event, also indicating the maximum level of water from heavy precipitation or flooding
- Iron deposits or the presence of reduced iron
- Hydrogen sulfide odor
- Sediment deposits resulting from water flow
- Drainage patterns, especially near streams, that show the maximum level of water that later receded back into the water body

Identifying the wetlands–nonwetlands boundary is critical, particularly in situations where real estate development is involved, since Section 404 of the Clean Water Act restricts the disturbance of regulated wetlands. Once identified, the boundary is marked in the field, sometimes with flagging (see Figure 9.7), which are then surveyed and marked out on a scaled map.

FIGURE 9.7
Field delineation of wetlands using flagging. (Courtesy of GZA GeoEnvironmental, Inc.)

The map is then used for planning purposes, which is usually subject to regulatory review.

Protective measures may be needed for restored and created wetlands, particularly in urbanized areas. Construction of a *buffer zone*, in the form of an undeveloped, vegetated band around the wetlands, or even a fence or other man-made barrier, can improve the odds of success for the wetlands and is, in fact, a requirement in many jurisdictions.

9.6 Threatened and Endangered Species Habitats

Of special consideration when assessing ecosystems are threatened and endangered species, which are protected under the Endangered Species Act of 1973 (see Chapter 2). The USFWS maintains databases of threatened and endangered plant and animal species at www.fws.gov/endangered. Most states maintain such databases of threatened and endangered plants and animals within their jurisdictions, which often include a significantly larger list of protected species that are only regionally rare, sometimes at the limits of the natural geographic ranges.

Spotting a threatened or endangered species during an ecological assessment is sufficient to document its existence, although spurious sightings of a specific animal can be ruled out if the data suggest that the animal is off course. The spotting of a threatened or endangered species may then trigger certain protections of the habitat of that plant or animal. However, understanding the breeding and nesting requirements, or life-cycle features of a threatened or endangered species is also critical to the protection of the species. For instance, constructing a dam on a river that a threatened or endangered species uses for spawning may put the species in jeopardy. It is not necessary to actually observe the fish in question if documentation exists that it has been observed in the river at some time in the past. Such observations indicate that the river is an endangered species habitat, triggering certain protections that may render the proposed dam construction project dead on arrival or too expensive to pursue due to the anticipated need for mitigation or modification in the project.

9.7 Ecological Mitigation and Restoration

9.7.1 Surface Water Restoration

Once the damage to a surface water body is understood, the next step is restoration of the water body. As with soil and groundwater remediation, it is important to establish goals to guide the restoration efforts. In general, the more humans interact with the surface water body, the more stringent the applicable standards.

The primary objective of the standards is to maintain the best and highest usage of the surface water body. Generic surface water cleanup criteria are usually based on exposure scenarios, using some of the following principles:

- Protected watershed requires the most stringent criteria.
- Freshwater surface water bodies require more protection than salt-water bodies.
- Other surface water categories, in descending order of importance, include drinking water sources; surface waters that sustain fish consumed by humans; surface waters that sustain only fish that are not consumed by humans; surface waters with no human consumption pattern but are consumed by native wildlife; and surface waters with no consumption by or habitat for humans or wildlife.

Even though generic standards are based on exposure scenarios, area-specific remediation standards can be generated by performing an ecological risk assessment, as described earlier in this chapter.

Step one in restoring a surface water body is to stop or significantly reduce the source of the contamination. When the contamination originates from a point source, such as a leaking underground storage tank, eliminating that point source is the beginning of the restoration process. In such cases, remediation may need to target only the impacted portion of the water body. Care must be taken to avoid damaging the water body in the course of fixing it, by inadvertently changing the water chemistry or the physical attributes of the water body in a way detrimental to the resident flora and fauna (as in the case of cultural eutrophication, as noted earlier in the chapter). Some plant communities may require special restoration, as described later in this chapter.

Sometimes the point source of the contamination, such as a storm sewer outfall, cannot be removed or the contamination originates from nonpoint sources. In such cases, the best way to restore the surface water body is by using the *total maximum discharge load* (TMDL) concept. Under the TMDL concept, stakeholders who live or work near the water body or discharge to the water body are apportioned TMDLs based on their operations and usage of the water body. The goal is to limit or decrease the aggregate contaminant load being introduced into the water body and allowing natural forces to clean up the water body over a period of time, in some cases years or even decades. Government often gets involved in forming stakeholder groups, assigning TMDLs, and monitoring and enforcing compliance.

9.7.2 Sediment Remediation

Dredging contaminated sediments from the surface water body can remove the main mass of contaminants. Dredging, however, has the potential of stirring up contaminated sediments and thereby risking the reintroduction of contaminants into the surface body's food chain, violating the "first, do no harm" principle. Hydraulic dredging is a process by which sediments are sucked through a pipe that trails at the bottom of the surface water body, almost like a vacuum cleaner. This method of dredging is more successful at avoiding the resuspension of contaminated sediments and is therefore more desirable from an environmental standpoint than traditional physical dredging, although it can be slower and more expensive than traditional dredging techniques.

9.7.3 Mitigation of Wetlands, Streams, and Aquatic Resources

When the impacts of a proposed action to wetlands, streams, or other aquatic environments are unavoidable, the USACE requires what is known as *compensatory mitigation*. The USACE (or approved state authority) usually determines the appropriate form and amount of compensatory mitigation required.

The types of mitigation generally fall into four categories:

- *Restoration*, which attempts to bring the wetlands back to its undamaged condition.
- *Enhancement*, which alters the wetlands to improve a particular function, such as support of a particular species, although that often occurs at the expense of other functions, such as support for other species.
- *Wetlands establishment*, which involves construction of wetlands where none ever existed. This can be accomplished only if wetlands conditions already exist at that location.
- *Preservation*, in which no actions are conducted that may damage the aquatic resource.

As described on the USACE Web site, compensatory mitigation is typically accomplished in one of three ways. One way, known as *permittee-responsible mitigation*, allows the entity seeking to perform the action to mitigate the wetlands through one of the aforementioned four methods. This type of mitigation may be provided at or adjacent to the impact site (i.e., on-site mitigation) or at another location, usually within the same watershed as the permitted impact (i.e., off-site mitigation). The permittee retains responsibility for the implementation and success of the mitigation project.

The other two methods of compensatory mitigation, mitigation banks and in-lieu fee mitigation, entail third-party involvement to achieve mitigation goals. They are forms of "third-party" compensation because a third party, the bank or in-lieu fee sponsor, assumes responsibility from the permittee for the implementation and success of the compensatory mitigation.

A *mitigation bank* contains wetlands, streams, and other aquatic resource areas that have been restored, established, enhanced, or preserved. A permit applicant can obtain credits to offset degradation of some other wetlands, stream, or aquatic resource area by buying credits from the bank. The value of the credits assigned to a mitigated area is determined by quantifying the aquatic resource functions restored, established, enhanced, or preserved.

A permit applicant can contribute to an *in-lieu fee mitigation* program, which will result in third-party restoration, creation, enhancement, or preservation activities of the aquatic resource. These programs are generally administered by governmental agencies or authorized nonprofit organizations.

Successful wetlands restoration must take into account proper site selection, hydrology, water quality, substrate augmentation and handling, plant material selection and handling, buffer zones placement, and long-term management. A brief overview of each element, as described in Kentula (1996), is presented next.

> *Site selection*—If a site was not formerly wetlands, it must either have the characteristics to become wetlands or such characteristics must be constructed. Existing wetlands in the same general area or a nearby area with similar land uses can be used to predict the success of a designed wetland.
>
> *Hydrologic analysis*—The desired plant community at a site must have the desired hydrology. If a wetlands has deteriorated due to sediment buildup and it no longer experiences year-round inundation, excavation to below the lowest anticipated water level within the water basin is warranted. The limits of hydric substrate along the basin banks will depend upon the slope of the banks. The lowest point of a properly constructed freshwater marsh, for instance, should be at a depth appropriate for emergent vegetation.
>
> *Water source and quality*—Mitigation measures for damaged wetlands, like surface water bodies, start with termination of the sources of pollution, which should have been identified during the assessment and delineation process.
>
> *Substrate augmentation and handling*—If hydric soils are not present, or have been damaged to the extent that they can no longer support wetlands vegetation, they will need to be augmented chemically or replaced. Augmenting, or mulching, the substrate with materials from a "donor" wetland enhances the substrate's ability to support wetlands plant communities by increasing water retention and decreasing soil loss and erosion. It can also introduce needed plant species, microbes, and invertebrates present in the "donor" wetlands

(although care needs to be taken to avoid the introduction of invasive species from the donor wetlands).

Plant material selection and handling—Establishing plant communities that resemble communities in similar, local wetlands will enhance the chances of success of the restoration project. Selecting species that establish themselves rapidly, can survive in a broad range of hydric and chemical conditions, and are not a preferred food of resident fauna will also improve the odds the project will succeed.

Buffer zone placement—The potential effects of nearby land uses, drainage patterns, and pollution sources will help in deciding the desired width of the buffer zone, although, as a rule, the wider the buffer zone, the better the protection of the wetlands. The width of the buffer zone also has to be weighed against the economic impact of loss of that land to development.

Long-term management—The most successful wetlands mitigation projects are self-sustaining, requiring no long-term management at all. In all cases, periodic inspections of the progress of the wetlands mitigation, involving the same steps as in an initial assessment (identification of plant communities, dominant species, soil types, hydrology, etc.) is warranted until the ecosystem appears to be healthy and stable.

References

Alabama Department of Environmental Management, www.adem.state.al.us/.

Barbour, Michael T., Gerritsen, Jeroen, Snyder, Blaine D., and Stribling, James B. 1999. Rapid Bioassessment Protocols for Use in Streams and Wadeable Rivers: Periphyton, Benthic Macroinvertebrates and Fish, 2nd ed. EPA 841-B-99-002. USEPA Office of Water.

Canter, Larry W. 1996. *Environmental Impact Assessment*, 2nd ed. New York: McGraw-Hill.

Federal Geographic Data Committee. July 2009. Wetlands Mapping Standard. FGDC-STD-015-2009.

Interagency Workgroup on Wetland Restoration. 2003. *Wetland Restoration, Creation, and Enhancement.* National Resources Conservation Service.

Kentula, Mary E. 1996. Wetland restoration and creation. In *National Water Summary on Wetland Resources,* Judy D. Fretwell, John S. Williams, and Phillip J. Redman, compilers, pp. 87–92. U.S. Geological Survey.

Lockheed-Martin Energy System, Inc. August 1997. Preliminary Remediation Goals for Ecological Endpoints. U.S. Department of Energy, Office of Environmental Management.

New Jersey Department of Environmental Protection. 2005. Field Sampling Procedures Manual.

New Jersey Department of Environmental Protection. 2008. Technical Requirements for Site Remediation, N.J.A.C. 7:26E.

New Jersey Department of Environmental Protection. August 2011. Ecological Evaluation Technical Guidance (draft).

Suter, Glenn W., II, Efroymson, Rebecca A., Sample, Bradley E., and Jones, Daniel S. 2000. *Ecological Risk Assessment for Contaminated Sites*. Boca Raton, FL: Lewis Publishers.

U.S. Army Corps of Engineers. January 1987. Corps of Engineers Wetlands Delineation Manual. Technical Report Y-87-1.

U.S. Environmental Protection Agency. August 1987. A Compendium of Superfund Field Operations Methods. EPA 540/P-87/001. Office of Emergency and Remedial Response.

U.S. Environmental Protection Agency. 1997. Ecological Risk Assessment Guidance for Superfund, Process for Designing and Conducting Ecological Risk Assessments. EPA 540-R-97-006. Office of Solid Waste and Emergency Response.

U.S. Environmental Protection Agency. 1998. Risk Assessment Guidance for Superfund, Volume II, Environmental Evaluation Manual. EPA/540/1-89/001.

U.S. Environmental Protection Agency. 2000. Bioaccumulation Testing and Interpretation for the Purpose of Sediment Quality Assessment, Status and Needs. EPA 823-R-00-001. USEPA Office of Water.

U.S. Environmental Protection Agency. April 10, 2008. Compensatory Mitigation for Losses of Aquatic Resources; Final Rule. 40 CFR Part 230.

U.S. Environmental Protection Agency. 2011. Polluted Runoff (Nonpoint Source Pollution), www.epa.gov/owow_keep/NPS/index.html.

U.S. Fish and Wildlife Service. 1979. Classification of Wetlands and Deep Water Habitats in the United States. FWS/OBS-79/31.

U.S. Fish and Wildlife Service, U.S. Environmental Protection Agency, U.S. Army Corps of Engineers, USDA Soil Conservation Service. 1989. Federal Manual for Identifying and Delineation Jurisdictional Wetlands.

U.S. Fish and Wildlife Service. National Wetlands Inventory, www.fws.gov/wetlands.

10

Environmental Impact Assessment and Mitigation

The *National Environmental Policy Act* (NEPA) requires that actions (such as construction projects) that may significantly impact the quality of the environment undergo a thorough evaluation of potential impacts. The NEPA process is triggered by any major federal action. There are also state and sometimes local NEPA equivalents of this process that are triggered by public and even private actions that exceed certain impact thresholds. The following is a brief list of projects that typically trigger the NEPA process:

- New bridges and tunnels
- New or major modifications to existing roadways or rail lines
- New or expanded airports
- Construction or relocation of government buildings and complexes
- Closure and reuse of military facilities
- New energy source exploration and distribution

The main thrust of NEPA is the evaluation of the environmental effects of the proposed action and its alternatives. The NEPA process is highly comprehensive in that it addresses socioeconomic and cultural resources in addition to natural resources that could be affected by a project. This chapter outlines the NEPA process and provides brief descriptions of the technical elements that are considered in the evaluation process.

10.1 The NEPA Process

At the beginning of the NEPA process, known as scoping, the public agencies with jurisdiction will meet and scope the project. One of the agencies involved is designated as the lead agency, which ultimately makes the decisions and through which all information flows. The lead agency defines the needs and objectives of a project, and manages the project from beginning to end.

10.1.1 Categorical Exclusion

In some cases, the government may issue a *categorical exclusion* (CATEX), which exempts the proposed action from the NEPA process. A CATEX is usually issued when a course of action is identical or very similar to a past course of action and the impacts on the environment from the previous action can be assumed for the proposed action, or when a proposed structure is within the footprint of an existing, larger facility or complex. The agency can then proceed with the project and skip the remaining steps to the NEPA process.

10.1.2 Environmental Assessment

If a CATEX is not issued, the process moves to the environmental evaluation step. Unless there are obvious potentially significant environmental impacts or substantial public concern about the project, the lead agency first will prepare an *environmental assessment* (EA).

Similar to a Phase I Environmental Site Assessment, (see Chapter 5), the EA evaluates potential impacts to the environment using existing data. However, the EA is much more diverse in scope than a Phase I Environmental Site Assessment. The types of data collected in support of an EA are discussed in Section 10.2. Potential impacts may be caused by "primary," that is, resulting directly from the construction and operation of the proposed project, or secondary, that is, resulting from the subsequent growth and development caused by the proposed project. There is no formal public participation requirement in the EA phase of the project (see Section 10.1.4).

If the EA indicates that no significant impact is likely, then the lead agency can release a *finding of no significant impact* (FONSI) and proceed with the project. Most EAs result in a FONSI, although the EA may suggest that it would be prudent to make some minor changes in the proposed project. In the instances where the EA indicates the potential of a significant environmental impact, an *environmental impact statement* (EIS) is prepared.

10.1.3 Environmental Impact Statement

The EIS is a tool for decision making, identifying potential impacts, and evaluating alternative actions that may reduce or eliminate the potential impacts. NEPA does not require that the actions cause no environmental impact at all, but rather that at the end of the process the *least environmentally damaging practicable alternative* (LEDPA) that complies with applicable federal, state, and local environmental laws is chosen and implemented.

An EIS typically begins with a statement of the purpose and need of the proposed action, and a description of the affected environment. This is followed by a listing of alternatives to the proposed action. For each environmental issue in each EIS, a *no action alternative* is evaluated, as is a range of alternatives. The no action alternative, in which the proposed action does not occur, is the baseline by which the other alternatives are compared. A comprehensive analysis of the environmental impacts of each of the possible alternatives is performed, as are cost analyses for each alternative, including costs to mitigate expected impacts. Oftentimes a cost-benefit analysis proposed action is conducted.

The first step in the preparation of the EIS is the *draft EIS* (DEIS). The DEIS fully describes the affected environment, provides a range of alternatives, and analyzes the expected impacts of each alternative. The DEIS also lists the permits, licenses, and so forth, needed to implement the project. Stakeholders are given the opportunity to comment on the DEIS during a specified comment period and a public hearing(s) is conducted during this period. The lead agency may respond to public comments by modifying the proposed action, developing new alternatives, modifying the analyses presented in the DEIS, or refuting stakeholder's objections.

After the lead agency addresses the comments on the DEIS, the document is finalized and a *final EIS* (FEIS) is issued. The FEIS describes the proposed action and its expected impacts on the environment. If stakeholders disagree with the FEIS, they can protest the decision to the director of the lead agency. Any resulting significant change in scope or alternatives will result in the preparation of a *supplemental EIS* (SEIS), beginning the NEPA process again or cancelling the proposed action outright.

Once all the protests are resolved, the lead agency issues a *record of decision* (ROD), which is its final action prior to implementing the proposed action. In the ROD, the lead agency issues its decision and discusses the findings of the EIS, demonstrating how its consideration of alternatives were incorporated into the agency's decision-making process. If members of the public are still dissatisfied with the outcome, they may sue the agency in federal court.

If the proposed action is expected to cause significant environmental impacts, an *environmental mitigation plan* is prepared and incorporated into the EIS documents and the ROD. Mitigation could include one or more of the following actions:

- Avoid the impact by changing all or part of a proposed action
- Limit the degree or magnitude of the proposed action
- Develop a plan to restore the impacted portion of the environment following implementation of the proposed action
- Develop a plan to restore the impacted portion of the environment over time
- Compensate for the environmental impact by replacing or providing similar environmental conditions elsewhere

10.1.4 Public Participation

The NEPA process involves the public and other stakeholders so that the current and useful information is available to the decision makers and conflicts with disgruntled stakeholders can be avoided. The system is designed so that citizens can be partners in the process.

During the NEPA process, the lead agency often will coordinate with stakeholders in an attempt to resolve any conflicts that may arise. The lead agency will develop a public participation program to inform the public and solicit participation in the decision-making process. The program must include notices in the *Federal Register* (once it has moved into the EIS stage) and also may include notices in local media, public hearings, and development of project-specific Web pages. Citizens and groups are asked to provide comments to the lead agency, which then may address the comments or incorporate them into the EIS. The lead agency frequently asks the EIS development team to provide technical input to the public participation program.

10.2 Technical Evaluation for the EIS

Extensive research goes into the preparation of an EIS. An EIS often takes years to prepare and may eventually consist of multiple volumes that can be thousands of pages in length. The EIS must be written for two audiences: the general public ("Joe Six Pack") and the regulatory technical experts. To respond to this challenge, the main body of the EIS is often written in layperson's language with technical information provided in the appendices. At a minimum, the following elements must be evaluated in the EIS (and the EA) with the level of analyses commensurate with the magnitude of the expected impact:

- Geology, soils, hydrogeology, and physiography
- Biology
- Surface waters
- Hazardous waste issues
- Air quality
- Noise
- Historic, archaeological, and cultural resources
- Transportation
- Socioeconomics

Each of these main categories is discussed in detail in the following sections.

Other elements that may need to be evaluated, depending on the type of project and the comments received during the scoping process, include light, energy, utilities, public health and safety, and solid waste and recycling.

10.2.1 Geology, Soils, Hydrogeology, and Physiography

Since the subject project involves construction, understanding of the soils, geology, hydrogeology, and physiography of the affected area is critical to the success of the project. Physiographic data are collected and evaluated, including topography, surface slopes, drainage patterns, and rates of erosion. Data are obtained regarding, for example, soil types, the physical characteristics of the soils, and the aquifers in the area. The underlying bedrock must be evaluated for its suitability to support the structures to be constructed and its hydrogeological aspects as well. Seismicity is usually evaluated, even in areas that are not earthquake prone.

Special attention is paid to groundwater in the NEPA process if the project area is located over a sole-source aquifer, that is, an aquifer that is the only major source of drinking water to the area. Also of concern is the impact of the proposed construction on groundwater recharge areas, which are the surface areas through which rain will penetrate into the subsurface and replenish the drinking water aquifer.

10.2.2 Biology

Evaluation of the biological setting of the project entails many of the activities described in Chapter 9. Of concern is the potential impact on flora (both upland and wetland) and fauna (both land animals and aquatic organisms).

Existing information often is available from sources such as the U.S. Fish and Wildlife Service, the U.S. Army Corps of Engineers (especially regarding wetlands), state environmental agencies, private environmental groups, and universities. Projects that may impact wetlands or endangered species habitats usually require an EIS and therefore the collection of field data that will help evaluate the potential impacts. Since the presence of flora and fauna vary seasonally in most parts of the country, field data may need to be collected in different seasons, which should be taken into account when scheduling the EIS.

Two classes of flora are to be considered: upland plant communities and wetland plant communities (as applicable). Upland plants (i.e., plants not associated with wetlands or water bodies) may be mapped by remote means, such as using aerial photographs, or by direct field inspections. Dominant plant species are identified in each plant community, as are any endangered plants or plants with some other special status. Wetlands delineation activities may be warranted if there is a question on the extent of jurisdictional wetlands in the project area.

Understanding the wildlife in an area usually entails identifying the presence of specific species, their relative abundance, and their habitats. This is especially important if endangered species or critical habitats may be present in the project area. Identifying habitats includes identifying the species' food (plant or prey), preferred vegetation cover for shelter or breeding, and territory size. Information on species habitat is usually presented in tabular form by species. Table 10.1 provides an example of such a table for an EIS prepared because of the proposed construction of a roadway in Seattle, Washington. That EIS also included similar tables for birds found in urban Seattle, marine mammals, and waterfowl.

Groups of aquatic organisms in or adjacent to the project area that may require study in the EIS may include benthic organisms (organisms that live at the bottom of the water body) that are the primary food source for most fish species, algae, aquatic plants, zooplankton, fish, and mammalian aquatic animals. Field activities include collecting samples from each aquatic community, from different depths in the water body in the case of zooplankton and fish. In most cases, a four-season study is warranted due to the wide variation in aquatic communities and life stages in the course of a calendar year.

Once species and their habitats are identified, the next step is assessing the potential short-term and long-term impacts of the proposed project on wildlife in the area using a variety of descriptive and statistical techniques.

TABLE 10.1

Land Mammals That May Be Found within Urban Habitat along the Alaskan Way Viaduct Corridor

Common Name	Scientific Name	Common Name	Scientific Name
Common opossum	*Didelphis marsupialis*	Muskrat	*Ondatra zibethicus*
Little brown myotis	*Myotis lucifugus*	House mouse	*Mus musculus*
Yuma myotis	*Myotis yumanensis*	Pacific jumping mouse	*Zapus trimtatus*
California myotis	*Myotis californicus*	Norway rat	*Rattus norvegicus*
Silver-haired bat	*Lasionycteris noctivagans*	Black rat	*Rattus rattus*
Big brown bat	*Eptesicus fuscus*	Coyote	*Canis latrans*
Hoary bat	*Lasiurus cinereus*	Raccoon	*Procyon lotor*
Townsend's big-eared bat	*Plecotus townsendii*	Ermine	*Mustela erminea*
Long-eared myotis	*Myotis evotis*	Mink	*Mustela vison*
Domestic rabbit	*Oryctolagus cuniculus*	River otter	*Lutra canadensis*
Eastern gray squirrel	*Sciurus carolinensis*	Domestic dog	*Canis familiaris*
Deer mouse	*Peromyscus maniculatus*	Domestic cat	*Felis domesticus*

10.2.3 Surface Waters

The first step in evaluating the potential impact of the proposed project on surface waters is to conduct a baseline evaluation of the surface water environment in the project area. This includes a detailed description of the surface water bodies; their relationship to drinking water supplies; a description of water flow, including wastewater effluent contributions to the surface waters; drainage patterns; and area floodplains.

New construction can affect water flow in many ways. Withdrawing water from a water body will reduce its availability to other users and may increase the relative concentration of pollutants in that portion of the water body. Conversely, addition of a new point discharge into a water body as a result of the construction project may increase the water body's flood potential and the flood potential of the receiving water body, in the case of a river, especially if there are floodplains in the study area.

Historic water flow data combined with computer simulations of water usage in the various alternatives under consideration aids in the identification of potentially significant environmental impacts to surface waters. Computer simulations also assess the potential changes in water quality as well from an increased load of pollutants and sediments into the water body.

Field sampling is used to fill in any data gaps that may be identified in the baseline study. Sampling parameters and methodologies are discussed in Chapter 9.

10.2.4 Hazardous Waste Issues

Understanding the who, what, and where of potential soil and groundwater contamination is critical in the planning stage. The procedures conducted in obtaining data about the natural environment in the area are similar to the research performed in support of a Phase I ESA: review existing documentation, analyze aerial photographs, and so on. Refer to Chapter 5 for an in-depth discussion on the evaluation of hazardous waste issues on real estate.

10.2.5 Air Quality

To understand the impact of the proposed project on air quality, a thorough understanding of the climate in the project area is needed. Such data are readily available from the National Oceanic and Atmospheric Administration (NOAA), the National Climatic Center, U.S. Environmental Protection Agency (USEPA), and state and local sources. Climate data that contribute to the EIS include monthly precipitation, wind speed and direction, mean monthly temperatures, daily temperature range, mean snowfall (as applicable), and heating/cooling degree days.

In addition, at a minimum, the changes in concentration in ambient air for air pollutants that are subject to the National Ambient Air Quality Standards (NAAQS; see Chapter 2) are considered in an EIS:

- Carbon monoxide and ozone precursors (NO_x and VOC), both which primarily are emitted by motor vehicle emissions
- Particulate matter (PM_{10} or $PM_{2.5}$), which is particulate matter associated with stationary energy sources (such as power plants) and diesel combustion engines (usually buses and heavy trucks)
- Total suspended particulates
- Sulfur dioxide, which is emitted primarily by stationary sources

Once baseline conditions are understood for these pollutants, impact calculations are performed to understand how they will change as a result of the project. The goal is the prevention of significant deterioration (PSD), as required under the Clean Air Act. This is done using computer modeling. Computer models incorporate the meteorological data, the baseline pollutant data, and the predicted changes in stationary and motor vehicle pollution sources to predict future increases in pollutants. They can estimate pollutant levels under normal conditions as well as worst-case conditions for the range of alternatives under consideration. If the computer model indicates that the project would not be in compliance with Clean Air Act requirements, then an optimization analysis would be performed, in which certain parts of the project would be changed and the resultant pollutant concentrations calculated. If the project causes significant increases in air pollution, it may result in a cancellation of the project altogether, especially if the project is located in a nonattainment area for that particular pollutant.

Separate from the study of NAAQS pollutants is the study of fugitive dust emissions during construction. Here as well, computer modeling can be used to calculate expected and worst-case scenario dust concentrations downwind of the proposed construction. Anticipated construction methods may need to be changed. Minor changes might entail the implementation of dust suppression techniques, such as air misting. More substantial procedural changes could result in a significant increase in the cost of the project.

Due to climate change concerns, many federal and state regulatory agencies are now requiring an analysis of greenhouse gas (GHG) emissions during the NEPA process. The primary GHG pollutants are carbon dioxide (CO_2), methane, nitrous oxide and fluorinated gases, with CO_2 typically being of primary focus. This evaluation consists of an estimate of the total GHG emissions from the construction and operation of the proposed action. If there is a predicted net increase in GHG emissions, then the EIS must present measures to attempt to offset this increase.

TABLE 10.2

Typical Outdoor Noise Levels in Various Environments

Setting	L_{dn} (dBA)
City noise (major downtown metropolis)	75–85
Very noisy urban	70
Noisy urban	65
Urban	60
Suburban	55
Small town and quiet suburban	40–50

Note: L_{dn}, day/night sound level; dBA, A-weighted decibel.

10.2.6 Noise

To understand the impact a proposed action will have on noise levels, it is first necessary to understand current noise levels. Whereas the decibel (dB) is the unit of measurement for noise, noise studies measure *A-weighted noise decibel* (dBA) levels. A dBA measurement put different weights on high-pitched and low-pitched noises since the human ear hears these sounds differently. Table 10.2 provides typical outdoor noise levels in urban, suburban, and rural environments. As with decibels, the dBA scale is logarithmic, with a sound level of 70 dBA being twice as loud as a sound level of 60 dBA.

Projecting noise levels from primary and secondary effects of the proposed project requires defining the most important noise sources and estimating the strength in the course of a day. For motor vehicle traffic, a major source of noise in a road construction project, traffic patterns must be understood and modeled. Various statistical descriptors are used to profile noise levels. For instance, L_{10} is the sound level that is exceeded 10% of the time, and L_{eq} is the constant, average sound level.

Measuring noise in the field is done using a simple handheld instrument. Once field measurements are collected, predicted noise levels of the proposed action are obtained, usually from data collected from similar projects. These data can be modeled using computer software and the results compared to regulatory noise criteria.

10.2.7 Historic, Archaeological, and Cultural Resources

Historic, archaeological, and cultural sites are properties or buildings with national or local significance. Historic sites are buildings, structures, and sites that date to colonial times. Archaeological sites are usually Native American sites that predate the colonial era. Cultural resources include cemeteries, parks, trails, and other similar features that may or may not have historical or archaeological significance but contribute to the local culture.

General information on historic, archaeological, and cultural resources is available on the National Register of Historic Places and various local equivalents. "Windshield surveys" (surveys conducted while driving) are performed in the area of interest. Structures are judged on their date of construction; their architectural merit; and their importance to local, state, or national history.

Sometimes areas that may contain historical or archaeological artifacts can be identified using infrared aerial photography, by the field identification of areas that appear to have been disturbed by humans, or by evidence of historic remains. If research indicates that a site may contain buried historical artifacts or structures, an archaeological investigation is performed. The area of interest is gridded, and test excavations or test cores are installed at set locations in the grid. The archaeologist analyzes the excavated soils in an effort to identify historic artifacts.

If research indicates that a property to be affected by the proposed action is or should be considered a historic or cultural resource, then the lead agency notifies the state historic preservation officer (SHPO), who, based on the evidence presented by the lead agency, either denies the historic or cultural significance of the property or proposes its inclusion onto the National Register. If it is determined that the proposed action will have an adverse effect on the historically significant property, the project will need to be modified, usually by changing the proposed action, or the artifacts or structures recovered or moved from their original location and preserved elsewhere.

10.2.8 Transportation

Transportation patterns are almost always disrupted by proposed actions. Transportation studies may involve evaluations of motor vehicle, pedestrian, rail, and airport traffic; however, motor vehicles are most commonly studied. The evaluation begins with a baseline study of the traffic (see Figure 10.1) and parking in and around the project area. Traffic engineers study the usage, capacity, and adequacy of the existing infrastructure, and compare it to accepted federal and state standards.

Computer models are used to estimate traffic demand after construction of the proposed project, as well as the impact on transit access and pedestrian routes from primary and secondary impacts. Models are run for the various alternatives for average traffic as well as peak times. Proposed traffic scenarios often must jibe with regional or local transportation master plans. They also interact with air quality because of the involvement of motor vehicles in the analysis.

Efforts to relieve significant impacts on traffic may result in adding, eliminating, or moving proposed entrance and exit ramps; changing the proposed design of arterial intersections; or adding, eliminating, or changing proposed traffic control measures, such as traffic lights, speed limits, and stop signs. Providing alternative modes of transportation may also be considered.

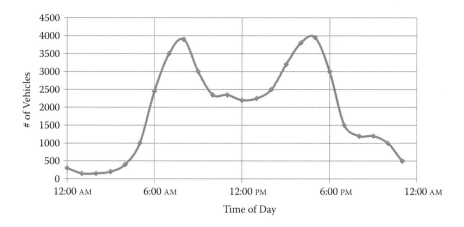

FIGURE 10.1
A typical highway traffic pattern over a 24-hour period. (Adapted from U.S. Department of Transportation, Federal Highway Division, July 2011, SR 99: Alaskan Way Viaduct Replacement Project: Final Environmental Impact Statement and Section 4(f) Evaluation.)

10.2.9 Socioeconomics

A major project can affect the socioeconomic conditions in the nearby local communities. Social impacts include the consequences of the proposed action that can alter the ways in which people live, work, play, and generally cope as members of society. In many cases, the secondary impacts of the proposed action are greater than the primary impacts. The study of the socioeconomic impact of the proposed action is perhaps the most subjective one within the EIS. The following socioeconomic categories must be considered in the EIS:

Demographics—The age, ethnicity, income profile, and level of poverty in the study area can all change due to the proposed project. Projected changes in demographics are estimated using statistical techniques.

Economic base—The effect of the proposed action on the local economy needs to be understood. Current economic data are available from federal and local economic development agencies, chambers of commerce, and so forth. Direct changes in the economy from the proposed project include the creation of construction jobs. Improved infrastructure may lead to more opportunities for local businesses, but businesses may also move in from outside the area, which could result in the closing or dislocation of local businesses.

Housing—Improvements in the local economy may result in an increase in housing prices, which could have a deleterious effect on middle-income or low-income residents.

Local government finances—The proposed action generally increases the economic base of the local government, but it also increases the long-term costs of governmental services needed to maintain the project. The resulting strain on public services, such as police, fire, schools, and health services, need to be taken into account when judging the long-term impacts of the proposed action on the local communities.

Land use—The effect of the project on land use needs to be understood, especially in its potential effects on open space and recreational properties, such as parks.

Aesthetics—The aesthetics of the proposed action may have a direct effect on the local communities. Significant efforts are made on the visual quality of the proposed action. Renderings are generally included in the EIS. There are usually numerous comments on the aesthetics of the proposed action, and many changes are made in response to public comments.

Environmental justice has moved to the forefront of socioeconomic concerns within the EIS. Environmental justice deals with the concept that the impacts of construction and industry often have a disproportionately high impact on low-income and minority communities. Outreach to these communities within and around the proposed action is conducted, and both primary and secondary effects of the proposed action on these communities are taken into account.

10.3 Environmental Consultant's Role in National Environmental Policy Act Process

Because the preparation of the EIS usually requires resources and expertise beyond the ability of the lead agency, it is usually prepared by a private-sector consulting firm. Since its preparation involves a number of disciplines, a large, diversified engineering firm generally leads the charge and forms a team of consultants. One of these consulting firms on the project team is usually an environmental consulting firm.

Environmental consulting firms provide input on many of the technical issues associated with the EIS, especially those dealing with land use, hazardous materials, and wildlife. Their scope includes preliminary research as part of the EA or EIS that is used to establish baseline conditions, and the collection of field data to fill in data gaps in the baseline data, or to assess the potential impacts of the various alternatives listed in the EIS. Computer modeling of air impacts, noise impacts, and so forth is also within the purview of environmental consultants. They may prepare fact sheets for stakeholders, attend public meetings as part of the evaluation team, and assist in the evaluation of alternatives and the redesign of the proposed project as warranted.

References

Bregman, Jacob I. 1999. *Environmental Impact Statements*, 2nd ed. Boca Raton, FL: Lewis Publishers

Council on Environmental Quality. December 2007. A Citizen's Guide to the NEPA: Having Your Voice Heard. Executive Office of the President.

Lawrence, David P. 2003. *Environmental Impact Assessment: Practical Solutions to Recurrent Problems*. New York: Wiley-Interscience.

U.S. Department of Transportation, Federal Highway Division. July 2011. SR 99: Alaskan Way Viaduct Replacement Project: Final Environmental Impact Statement and Section 4(f) Evaluation.

U.S. Environmental Protection Agency, www.epa.gov.

U.S. Environmental Protection Agency. 1974. Information on Levels of Environmental Noise Requisite to Protect Public Health and Welfare with an Adequate Margin of Safety. Report Number 550/9-74-004.

U.S. General Services Administration, Laws and Regulations, www.gsa.gov/portal/category/21006.

Section IV

Indoor Environmental Concerns

11

Asbestos Surveying

11.1 Introduction

It was known as the Magic Mineral. Asbestos, a naturally occurring mineral that was valued since ancient times for its resistance to heat and flame, became one of the most prevalent and vexing environmental problems in recent times. Its long history goes back to the ancient Greeks, who gave the mineral its name, which means "inextinguishable." Its modern history began in the mid-1800s, when the construction industry began incorporating asbestos fibers in its roofing materials to prevent fires from spreading in urban areas.

During the 1900s, asbestos usage in building materials grew due not only to its fire-resistance abilities but also its ability to provide insulation, sound-dampening, tensile strength, and general abundance. It was even used for food processing and by the motion picture industry as artificial snow (e.g., in the *The Wizard of Oz*). There was a large increase in usage during World War II as an insulation material for naval vessels. After World War II, it was incorporated into a huge array of building materials and other materials. To this day, it is still used in various roofing products and in vehicular break linings and other nonbuilding materials.

11.1.1 Types of Asbestiforms

There are six known asbestos minerals, as follows:

- *Chrysotile*, a white or greenish-colored mineral, was the most commonly used asbestiform in the United States (see Figure 11.1).
- *Amosite*, also known as brown asbestos, is an amphibole asbestiform that derives its name from the Asbestos mines of South Africa.
- *Crocidolite*, also known as blue asbestos, was mainly mined in South Africa and Australia, and had limited usage in the United States.
- *Anthophyllite* has excellent resistance to chemicals and heat, and was primarily used in the United States in decorative and acoustical material.

FIGURE 11.1
Naturally occurring chrysotile asbestos. (Courtesy of the Natural History Museum of London.)

- *Tremolite* is a white- to yellow-colored asbestiform that has a major ingredient in industrial and commercial talc. There is also a non-asbestos form of tremolite.
- *Actinolite*, which is greenish to white in color, has poor resistance to chemicals and had limited commercial usage. It may be found in commercial and industrial talcs.

Except for chrysotile, all of the asbestiforms fall into the mineralogical category of amphiboles, which form hard, needlelike particles that will stick to and damage soft tissues in the lungs and other internal parts of the body. Alternatively, as its name implies, serpentines are categorized by their wavy morphology, tending to get lodged predominantly in the upper respiratory pathways. Amphibole asbestiforms are generally recognized as more hazardous to human health than chrysotile.

11.1.2 Health Problems Related to Asbestos

Asbestos-related health problems arose as asbestos usage grew in popularity. In fact, Henry Ward Johns, an asbestos pioneer and founder of the company that was forerunner to the Johns-Manville Corporation, died of an asbestos-related disease in the late 19th century, a victim of his own successful business venture.

There are three known asbestos-related diseases: asbestosis, lung cancer, and mesothelioma.

Asbestosis is a scarring (fibrosis) of the lung. Scarring in the alveoli of the lung impairs the elasticity of the lung tissue and hampers its ability to exchange gases. The irreversible disease restricts breathing, with a latency period of 10 to 30 years. It is generally associated with a long, heavy exposure to airborne asbestos fibers.

Lung cancer is a malignant tumor of the bronchi covering. Although there are many causes of lung cancer, there is a clear increase in risk among people who worked with asbestos. Moreover, there is no threshold or limit of exposure below which the risk of lung cancer is not increased. The typical latency period for lung cancer resulting from asbestos exposure is 20 to 30 years.

Mesothelioma is a cancer of the mesothelium, the lining of the chest or the lining of the abdominal wall. It is the only known type of cancer directly attributed to asbestos exposure. Unlike most other cancers, by the time it is diagnosed, it is almost always fatal. There is no exposure threshold for mesothelioma, and the disease may not manifest itself until up to 40 years after the time of exposure to asbestos.

11.1.3 Regulatory History

The first major federal regulation regarding asbestos in buildings came about in 1972, when the Occupational Safety and Health Administration (OSHA) limited the concentration of airborne asbestos in the workplace at 5 fibers per cubic centimeter (f/cc). Over the years, that action level has been progressively lowered to the current permissible exposure limit (PEL) of 0.1 f/cc.

In April 1973, the United States Environmental Protection Agency (USEPA) banned the application of asbestos-containing spray-on surfacing material and visible emissions of asbestos under the National Emission Standards for Hazardous Air Pollutants (NESHAP) section of the Clean Air Act. Other forms of asbestos-containing materials (ACMs), which the USEPA defines as having greater than 1% asbestos by weight, were gradually phased out over a number of years.

The regulation with perhaps the greatest impact on the world of asbestos was the portion of the Toxic Substance Control Act (TSCA) that became known as the Asbestos Hazard Emergency Response Act (AHERA) of 1986. AHERA, which was designed to minimize asbestos hazards in schools for grades K through 12, set up elaborate systems for the testing, reporting, training, and maintenance of asbestos-containing building materials inside schools, which AHERA coined as "ACBM," inside buildings on school grounds. ACBM, as defined under AHERA, did not include exterior materials, such as roofing materials and exterior wall coverings, as well as temporary equipment, such as laboratory hoods, blackboards, and other laboratory equipment.

AHERA set up several categories for people licensed to handle ACBM. An *AHERA inspector* is an individual who is appropriately trained to conduct an asbestos survey, which involves identifying suspect ACBM, and often

collecting bulk samples of suspect material for laboratory analysis to identify and quantify the presence of asbestos fibers in the material. Contrary to
popular belief, an asbestos survey does not necessarily involve the collection
of bulk samples, as explained in Section 11.3. An *AHERA management planner* is an individual who can assess the hazard posed by ACBM in a school.
Very often the same person is the AHERA inspector and management planner. The other title set up under AHERA is the *AHERA project designer*, who
designs the actions to be undertaken to abate the asbestos hazard identified
by the management planner.

When AHERA was promulgated, it predicted a modest economic impact
on the newly regulated local education agencies (LEAs), whose job it is to
implement the AHERA regulations:

> The cost of an asbestos inspection is estimated to range from $1,144 to
> $1,627 per school for schools with both surfacing and thermal system
> insulation ACM … assuming the average school has to analyze 20 sam-
> ples, the cost of analysis will be $500 to $940 per school.

The actual price was significantly higher, as billions of dollars were spent
on abating asbestos hazards in the schools. The methodologies prescribed
under AHERA soon became the industry standard by which other methodologies are compared.

ASHARA, or the Asbestos School Hazard Abatement Reauthorization Act,
was enacted in 1990 and took effect in 1994. It extends the AHERA rule to
nonschool buildings that are not private residences with less than 10 dwelling
units. Through ASHARA, the AHERA regulation became required in virtually all commercial properties. The following subsections describe the inspection and hazard assessment techniques prescribed under AHERA.

11.1.4 Types of Asbestos-Containing Materials

As indicated earlier, asbestos was used in literally thousands of products. A
summary of building products in which asbestos was used is provided in
Table 11.1.

Certain materials are allowed to be designated as non-ACBM per se
(without the need for sampling or obtaining corroborating documentation),
including the following:

- Poured concrete, concrete (cinder) block, bricks, and terra cotta materials
- Glass products
- Unpainted wood
- Fiberglass insulation—Fiberglass is thermal insulation that more-
  or-less universally replaced asbestos. It is typically yellow-colored,
  and much more "squeezable" than ACM thermal insulation

TABLE 11.1

Products That May Contain Asbestos

Cement asbestos insulating panels	Insulation, thermal sprayed-on	Cooling tower, fill
Cement asbestos wallboard	Blown-in insulation	Cooling tower, baffles or louvers
Cement asbestos siding	Insulation, fireproofing	Valve packing
Roofing, asphalt siding	Taping compounds	Waterproofing, asbestos base felt
Roofing, asphalt saturated asbestos felt	Packing or rope (at penetrations through floors or walls)	Waterproofing, asbestos finishing felt
Roof, paint	Paints	Waterproofing, flashing
Roofing, flashing (tar and felt)	Textured coatings	Damp proofing
Roofing, flashing (plastic cement for sheet metal work)	Flexible fabric joints (vibration dampening cloth)	Plumbing, piping insulation
Laboratory hoods	Fire curtains	Plumbing, pipe gaskets
Laboratory oven gaskets	Elevators, equipment panels	Plumbing, equipment insulation
Laboratory gloves	Elevators, brake shoes	Electrical ducts (cable chases)
Laboratory bench tops	Elevators, vinyl asbestos tile	Electrical panel partitions
Putty and/or caulk	HVAC piping insulation	Electrical cloth
Door insulation	HVAC gaskets	Insulation, wiring
Flooring, asphalt tile	Boiler block or wearing surface	Stage lighting
Flooring, vinyl asbestos tile	Breeching insulation	Incandescent recessed fixtures
Flooring vinyl sheet	Fire damper	Chalkboards
Flooring, backing	Duct insulation	Ceiling tile
Plaster, acoustical and decorative	Ductwork taping	Flue, seam taping

- Rubberized or plastic insulation materials
- Wallboard (Sheetrock)
- Metallic objects and sheet metal

11.1.5 Components of Buildings

To identify suspect ACMs in a building, it helps to understand the construction of buildings.

A building superstructure consists of a foundation load-supporting walls (i.e., walls designed to support all or part of a building) and a roof. Building foundations tend to be poured concrete and are therefore not considered to be suspect ACM. Exterior walls and interior load-bearing walls are typically composed of concrete (cinder) block, brick in older buildings, or, in taller structures, a steel I-beam frame. None of these materials, excepting the mortar used in the wall construction, are considered to be suspect ACM. Steel I-beams are often sprayed with fireproofing materials, especially indoor beams (see the description of "surfacing material," in Section 11.2).

Building improvements, on the other hand, are often composed of suspect materials. Foundations and floors are often covered by resilient floorings or carpeting, whose adhesive is a suspect ACM. The current material of choice in the construction of interior, nonload bearing walls are wallboard panels that are installed on aluminum floor and ceiling runners. The joint compound used to smooth out the seam between the panels is a suspect ACM. Older buildings have interior walls composed of plaster, which is a suspect ACM. The plaster was usually mixed and poured on site onto a metal lath or wood lattice to give it shape, or directly onto the brick or cinder block superstructure. It also usually has a base coat and a different-colored top coat, both of which are considered to be separate materials for asbestos testing purposes. Before being banned, exterior walls were often covered with durable transite panels or transite shingles. Nowadays, exterior walls of commercial buildings are often covered with decorative brick façades (buildings constructed in the United States in the last few decades were almost never constructed with structural brick).

Structural floors are often composed of concrete poured onto corrugated steel sheeting. The bottoms of these sheets, which are the ceilings of the floor below, are often sprayed with fireproofing material. Office buildings are usually equipped with suspended ceiling grids, often known as "drop" ceilings. These ceilings are comprised of an aluminum lattice with ceiling tiles that are designed to fit within the lattice spaces. These ceiling tiles are commonly 2' × 4' or 2' × 2' in area and come in a wide variety of styles and textures. Fibrous ceiling tiles are generally considered to be suspect ACM, unless they are composed of fiberglass. Nonfibrous ceiling tiles are often present in exterior areas, such as covered parking lots where moisture is an issue, or in food preparation areas, where fibrous materials are often banned to protect the food preparation process. In these areas, ceiling tiles are often composed of Sheetrock or plastic, and are therefore not suspect ACM.

Boiler rooms and other building operations rooms can often be a cacophony of piping and wiring that can intimidate the novice building inspector. Basically, a building's piping is designed to convey heat or heating materials, water, or electricity. Prior to its banning, cementitious piping was usually composed of transite, and the piping insulation often contained asbestos. Therefore, it is essential for an inspector to identify the different piping systems in a building and be able to distinguish between them.

> *Heating pipes*—Pipes in a building may convey steam or hot water to heat the building, or natural gas or petroleum to supply the boiler with combustible fuel to heat the building. Steam and hot water

pipes are always insulated, both to prevent heat loss and to protect building occupants from burns. Natural gas pipes leading to a boiler are usually not insulated, but fuel oil piping can be, especially if the oil has been preheated, as is the case for no. 6 fuel oil or bunker oil (see Chapter 3).

Water pipes—Water pipes have three purposes: to convey heated water to baseboard radiators or stand-alone radiators for heating purposes; to convey drinking water to facility bathrooms, kitchens, and drinking fountains; or to convey wastewater from the building to a sewer or septic system. As mentioned, heated water pipes are almost always insulated, as are drinking water pipes, to prevent the water from warming to room temperature or from freezing in unheated rooms during northern winters. Water pipes are usually ½ inch to 2 inches in diameter, and are connected to water heaters at one end, and water dispensing or heating equipment at the other end. Waste water pipes are almost never insulated. They are typically wide pipes, usually 4 inches or more in diameter.

Electricity—Electrical conduit is almost never insulated. It is recognized by its narrow diameter and its connection to outlets and other electrical equipment.

To determine the usage of a pipe, the inspector may have to trace it through the building, across walls and floors, to either of its endpoints so that the liquid that it is conveying becomes apparent.

Many parts of a building's heating, ventilation, and air-conditioning (HVAC) systems contain fibers and are suspect ACM. They are described in Chapter 17.

11.2 Classifying Suspect Materials

To facilitate grouping of suspect ACBM into categories, a list of all suspect ACBM observed during the inspection should be developed. Groupings of suspect ACBM are called *homogeneous areas*, which are defined as materials that are alike in color and texture. Suspect ACBM similar in appearance but found in different areas of the building that were constructed at different times should be considered separate homogeneous areas, unless building records indicate contemporaneous application. Plant maintenance personnel or the property owner may provide insight in determining the application history of suspect ACBM. The inspector should group materials into separate homogeneous areas if no definitive determination can be made based on the available data.

Suspect ACBM must be divided into one of three categories, as defined under AHERA:

Surfacing material—This category includes materials that were applied with a sprayer, a trowel, or some similar piece of equipment. Typically, they are prepared for application on site. Spray-on insulation, typically applied to support structures, protects the steel from losing compressive strength in a fire. Figure 11.2 shows a so-called popcorn ceiling texture common in hotel rooms and office corridors. Applied with a trowel, this surfacing material is valued for its ability to reduce glare and deaden sounds.

Thermal system insulation—This category includes materials that are attached to pipes, fittings, boilers, breaching, tanks, ducts, or other components to prevent heat loss or gain or water condensation. Typically, these materials are manufactured off site and cut to size on site. The so-called "aircell" piping insulation shown in Figure 11.3 is always ACM; the asbestos was placed inside the honeycomb structure to provide insulation to an otherwise poorly insulated cardboard structure. The boiler jacketing shown in Figure 11.4 may be composed of asbestos, fiberglass, or some other synthetic material, and needs to be investigated as to its composition.

Miscellaneous ACBM—This category includes building materials not included in the first two categories. Miscellaneous materials inside a building include floor tiles, ceiling tiles (see Figure 11.5), various adhesives, and joint compound applied to wallboard. Exterior wall shingles and corrugated transite panels (see Figure 11.6), are cementitious, prefabricated products that were applied to the exterior of buildings. They were both highly valued for their durability. Roofs are often complex systems consisting of many layers and materials, many of which may contain asbestos, especially tars, known as "flashing," which are used as sealants around chimneys and other roof openings, and felt papers that underlie shingles and other resilient roof coverings, and are designed to prevent moisture intrusion.

There may be many types of miscellaneous material in a building that are not building components, such as theater curtains and lab tabletops. However, theater curtains are not a building material and therefore do not have to be assessed under the AHERA rule. The decision to address suspect nonbuilding material is up to the inspector and building owner. Documentation of all the suspect ACM found in the building is certainly the better route to follow but is optional.

Suspect ACBM can also be distinguished by their *friability*, which is the ability for the material to be pulverized using hand pressure. AHERA

FIGURE 11.2
Spray-on popcorn ceiling texture. (Courtesy of GZA GeoEnvironmental, Inc.)

FIGURE 11.3
Suspect ACM in the form of aircell piping insulation. (Courtesy of GZA GeoEnvironmental, Inc.)

FIGURE 11.4
Suspect ACM in the form of insulation wrap around a hot water tank. (Courtesy of GZA GeoEnvironmental, Inc.)

FIGURE 11.5
Suspect ACM in the form of lay-in ceiling tiles. (Courtesy of GZA GeoEnvironmental, Inc.)

FIGURE 11.6
Corrugated transite panels on a building exterior. (Courtesy of GZA GeoEnvironmental, Inc.)

requires the inspector to physically touch the suspect material to determine its friability.

AHERA developed the concept of a *functional space* to simplify the inspection and the grouping of homogeneous materials. Functional spaces are spatially distinct units within a building that sometimes contain different populations of building occupants. For example, a classroom is a functional space because it is enclosed and separate from the rest of the building. In the same vein, a boiler room is also its own functional space. Pipe chases, air shafts, elevator shafts, and return air plenums are also separate functional spaces, even though they are unoccupied.

AHERA allows considerable latitude for defining functional spaces. A long corridor could be divided into separate functional spaces if doing so is useful to identify important distinctions in the conditions or disturbance of ACBM. By allowing the inspector to isolate the problem area, the building owner has more options in determining response actions.

One homogeneous material may be found in several functional spaces. For example, sprayed-on ceiling material may cover an entire floor, including classrooms, hallways, and bathrooms. A few sites would be selected among all of the functional spaces for sampling purposes, and the sampling locations would be documented according to the functional space.

11.3 Performing the Asbestos Survey

The objective of an asbestos survey is to identify and locate all ACBM in the subject building. However, not all suspect ACBMs or ACMs need to be physically sampled for asbestos content. If a building was constructed before 1981, then the suspect materials in that building can be classified as presumed asbestos-containing materials (PACM) under OSHA. As PACM, they must be managed as ACM (or ACBM), and handled as ACM or ACBM if removed. So the minimum amount of bulk samples that need to be collected in an asbestos survey is zero.

However, in most cases a property owner, lender, or other interested party wants a fuller understanding of the asbestos hazard (and associated costs) in a building. To gain a fuller understanding of the presence of asbestos and asbestos hazards in a building, bulk samples must be collected of suspect materials. This section describes the general procedures for conducting an asbestos survey of a building.

11.3.1 Before the Inspection: Doing Your Homework

A little bit of research may go a long way in simplifying the asbestos survey. The asbestos inspector should review documentation that may be available regarding the construction history of the building. Construction documents, or contract documents, that were used to construct the building may consist of working drawings, specifications, addenda, change orders, and shop and as-built drawings. These documents and their addenda may indicate the locations of suspect ACM, or, in some cases, actually specify the brand and model number of building material to be used, which can then be cross-checked with the former manufacturer. Working drawings that were subject to change during the building construction can only be used for guidance purposes. Construction blueprints may be available on the property, or in the local or county building offices.

11.3.2 Building Walk-Through

The inspection should begin with a walk-through of the entire building. All interior areas must be inspected, including the space above suspended ceiling grids in each room, crawlspaces, tunnels, mechanical areas, basements, attics, and air plenums. The type, mount, and condition of suspect ACBM should be noted on the building drawings or plans. If drawings are not available at that time, the inspector should draw field sketches. In areas where physical access is impossible (e.g., pipe chases), the inspector should make an effort to determine the location and physical condition of the material. The inspection should include halls, closets, attic spaces, and tunnels, as well as walls, ceilings, beams, ducts, and any other surfaces. ACBM is sometimes found in areas that are deemed inaccessible (e.g., behind walls, false ceilings, etc.).

For each functional space, the inspector should prepare a diagram, approximately to scale, that shows all suspect ACBM in the space. Such a diagram is presented in Figure 11.7. If the functional space contains discontinuous areas

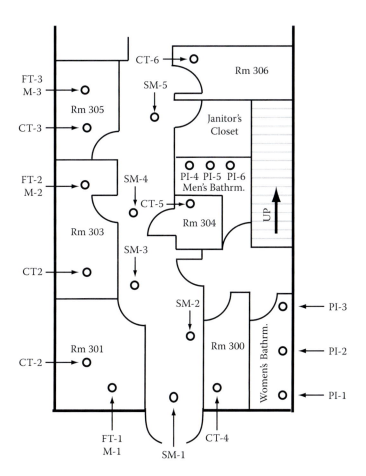

FIGURE 11.7
Scaled drawing of an office building showing the locations of bulk sample collection. In this diagram, the sample names are indicative of the material sampled. FT, floor tile; M, mastic; CT, ceiling tile; SM, surfacing material (in this case, popcorn ceiling texture); PI, pipe insulation.

of suspect ACBM (e.g., homogeneous areas on consecutive floors of a building), the inspector should sketch each separate area and place all sketches on the sample graph as close together as possible. The functional space may contain areas that are not in one plane, for example, a ceiling and a wall with the same type of suspect ACBM. In this case, the inspector should sketch each flat surface and place the sketches on the same graph as close together as possible.

11.3.3 Designing a Sampling and Analysis Plan

To confirm whether a material is ACBM, the inspector can either sample and analyze suspect material, or presume that the material is ACBM. Any building material can be designated as PACM; this designation is commonly applied to nonfriable materials or to materials that are small quantities. However, assuming that a suspect material is PACM may restrict building operations or increase the costs of future renovations or demolitions. Therefore, it is best to decide prior to the actual survey what categories of suspect materials are to be sampled and what categories of suspect materials are to be designated as PACM. If it is decided that a suspect ACBM is to be sampled, sampling locations must be selected.

Materials installed in different periods may appear to belong to the same homogeneous areas. If the function of the material is different, then they should be considered separate homogeneous areas (e.g., ceiling plaster, which may be cosmetic, versus wall plaster, which may be an acoustical insulation). If there is any reason to suspect that materials might be different, even though they appear uniform, they should be assigned to separate homogeneous areas. For example, materials in different wings of a building, on different floors, or in special areas such as cafeterias, machine shops, and band rooms might be assigned to separate homogeneous sampling areas, unless there is good reason to believe that the material is identical throughout.

In a large multifloor building (more than 10 stories), separate homogeneous area sampling for each floor may not be necessary. Material that appears identical on every floor can be grouped into one homogeneous sampling area, especially if these areas have been maintained by the same entity (this assumption usually cannot be made in multitenant apartment buildings or commercial condominiums, for example). When in doubt, the inspector should assign suspect materials to separate homogeneous sampling areas.

11.3.4 Sampling of Layered Materials

Some materials are manufactured or installed in layers. Plaster is typically installed in two layers: a brown undercoat and a white top coat for aesthetics. Since each layer was mixed and installed separately, they must be treated as separate homogeneous areas. However, from a practical standpoint, it is almost impossible to sample them separately in the field, since they are attached to each other and the white top coat is ordinarily very thin. What most practitioners do is send a composite sample to the laboratory and request that the laboratory separate the two materials and analyze them individually.

A similar procedure is usually practiced for layered roofs (see Figure 11.8). Most roofs consist of several layers, each layer performing a specific function. For instance, Figure 11.8 shows a layered roof system consisting of shingles (for durability), a roofing felt (which acts as a vapor barrier), and the rafter (for structural strength). The shingles and roofing felt must be treated as separate homogeneous areas. In some roofing systems, it is not uncommon to have several different layers of materials, and, in older buildings, multiple roofs is

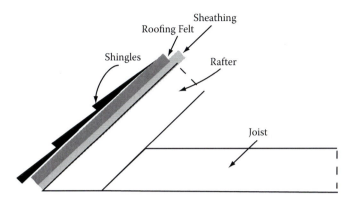

FIGURE 11.8
A schematic diagram of a typical layered roofing system.

not uncommon. Add in roof flashing and other roofing materials, and there can be numerous homogeneous areas of suspect ACM on a single roof.

Another common layered system is wallboard and joint compound, which is applied between wallboard sheets to smooth out the surface. Although wallboard generally is not considered to be a suspect material, joint compound commonly contained asbestos prior to its ban in the 1980s. As with plaster, it is often impossible for the inspector to separate the two materials in the field. Because the wallboard is not considered suspect ACM, a composite sample (wallboard + joint compound) almost always tests negative for ACM. In the early 1990s, the USEPA ruled that the joint compound must be analyzed separately from the wallboard. Separating the joint compound from the wallboard is commonly performed in the laboratory.

11.3.5 Sampling of Homogeneous Materials

For thermal system insulation (TSI), the inspector should collect at least three randomly distributed bulk samples from each non-PACM homogeneous area. At least one bulk sample should be collected from each homogeneous area of non-PACM patched thermal system insulation that is less than 6 linear or square feet. The samples should be collected in a manner sufficient to assess whether the material is ACBM from each non-PACM insulated mechanical system where cement is used on tees, elbows, or valves. Bulk samples are not required to be collected from any homogeneous area where the TSI is fiberglass, foam glass, rubber, or other non-ACBM.

The USEPA recommends collecting nine samples from each homogeneous sampling area of surfacing material. However, AHERA requires the following numbers in its sampling protocol:

Square Footage	Required Samples	Recommended Number
<1000 ft²	3	9
1000–5000 ft²	5	9
>5000 ft²	7	9

Choosing sample locations according to personal judgment produces samples that may not be representative and can lead to a wrong decision about the presence or absence of asbestos. Therefore, AHERA requires dividing the suspect surfacing material into a 3 × 3 grid. This can be done by eye; exact measurements are not needed. For greatest coverage, one sample from each of the nine regions should be collected. If fewer samples are to be collected, the inspector should establish a random sampling scheme to avoid biases from creeping into the sampling plan. The inspector should collect samples from approximately the center of a subarea or as close as possible to the center, if accessible, unless the presence of light fixtures and so forth makes sampling from the center location impractical. For very irregularly shaped areas, the sampling area may be divided into nine approximately equally sized subareas that do not necessarily form a rectangular grid.

AHERA recommends the collection of a minimum of three samples from each homogeneous area of miscellaneous, nonbuilding materials such as peg or cork boards, fabrics, laboratory vessels, transite panels, and fireproofing materials that are not PACM. Only one bulk sample is required of these materials, however.

Sampling suspect ACBM from inaccessible areas can only be accomplished by a partial or total destruction of building materials of the obstructing materials. Typical destructive procedures include removing cinder block, climbing along beams and rafters, and cutting holes through plaster or wallboard to access the suspect material. This procedure is typically performed only during predemolition or prerenovation surveys, when future partial or complete demolition is expected.

11.3.6 Bulk Sampling Procedures

To minimize damage to ACBM and the dangers of possible fiber release, samples should be collected in an unoccupied area when possible. The inspector should wear at least a half-face respirator with disposable high-efficiency particulate air (HEPA) filters, unless a higher protection factor is necessary or desired. The inspector should then perform the following steps to collect the bulk sample:

1. Wet the surface of the material to be sampled with "amended water" (water with a glycol additive to soften the material) mist from a spray bottle or place a plastic bag around the sampler with the open end of the bag pressed tightly against the wall or ceiling.

2. Sample with a cork borer or knife. Be sure to penetrate any paint or protective coating and all layers of the material. Decontaminate the sampling device if it is to be used again.

3. Place the bulk sample into a sample container. Avoid any container that may break, tear, or lose its lid if accidentally dropped. Label the container with the date and unique sample ID number.

4. Clean debris using wet towels or sponges and discard them into a plastic bag.

5. Use a sealant or a bridging material such as plaster or other enclosure material such as duct tape to cover the spot where the sample was extracted. In the case of thermal system insulation, damage may be repaired by applying an encapsulant and wrapping with duct tape. The building owner should be notified if further repair work is deemed necessary.

6. Double-bag all samples by placing them in a larger zip-lock bag or sealed plastic bag before transporting.

11.3.7 Quality Assurance Program

A good quality assurance program will ensure the reliability of the laboratory's analytical results. One type of quality control (QC) sample, known as a *duplicate sample*, involves collecting two samples immediately adjacent to each other ("side-by-side" samples), or splitting one sample into two separate containers. The QC samples are labeled and handled in the same way as ordinary samples, except that it may be prudent to use different designations to disguise the fact that the samples should yield identical results. The inspector should collect at least one QC sample per building or one QC sampler per 20 samples, whichever is larger. In some cases, the inspector may send the QC sample to a second laboratory to confirm the results of the first laboratory. Any disagreements about the presence or absence of asbestos should be investigated by reanalyzing the samples or collecting additional samples.

11.4 Laboratory Analysis of Bulk Samples

Selection of an analytical laboratory should be based on successful participation in a national or state-sponsored quality assurance program. American Industrial Hygiene Association (AIHA) accreditation is also a desirable factor when choosing an analytical laboratory.

The recommended method for the preliminary analysis of bulk samples employs the use of polarized light microscopy (PLM) with dispersion staining and stereobinocular examination. This analytical technique is based on the optical properties of crystalline and noncrystalline substances. It is a quick method for estimating asbestos content in a bulk sample. Figure 11.9 shows the appearance of chrysotile fibers under a polarizing microscope.

The major disadvantages to the PLM method are:

- Because it is qualitative in nature, there can be large variations in estimates of asbestos content, depending on the microscopist.
- It has difficulty detecting extremely small amounts of asbestos or short fibers.
- The lack of resolution when the asbestos fibers are bound in an organic matrix. Building materials with organic binders include vinyl floor tiles (VFT), adhesives, roofing materials, asphalt shingles, and caulks.

Point counting is recommended when a bulk sample is found to contain less than 10% asbestos fibers using qualitative PLM analysis and mandatory in certain jurisdictions. Under the point counting method, the visual field under a microscope is divided into a 10 × 10 grid. The microscopist observes each of the 100 squares in the field of vision and records the presence or absence of asbestos fibers in that square. By tallying the square with asbestos

FIGURE 11.9
Asbestos fibers as seen under polarized light microscopy. (Courtesy of Materials and Chemistry Laboratory, Inc., Oak Ridge, Tennessee, www.mcl-inc.com.)

fibers, the microscopist generates a quantitative, more reliable estimate of the percentage of asbestos fibers in the bulk sample.

Because of the difficulty in obtaining reliable readings using PLM on organically bound materials, the use of the nonfriable organically bound (NOB) protocol for such substances is recommended and in some jurisdictions required when PLM has not identified the suspect material as ACBM. The NOB method involves applying acid to the material to dissolve the organic matrix, and then burning the material. The residue will contain ash and asbestos fibers. If the residue is less than 1% of the original sample, no further analysis is required and the suspect material can be reported as non-ACBM. If not, the residue must be analyzed by PLM or by the more accurate yet more expensive transmission electron microscopy (TEM). TEM is also more reliable in identifying short or thin asbestos fibers.

11.5 Interpretation of Results

If one or more samples from a homogeneous area contains more than 1% asbestos, then the inspector is required to treat the entire homogeneous area as ACBM. To save money, inspectors generally use what is known as the "positive stop method." As its name suggests, the inspector will instruct the laboratory to stop the analysis of the remaining samples from a homogeneous area once one of the samples is shown to be ACBM. This approach is justifiable since AHERA requires a material to be designated as ACBM if there is one positive reading from the homogeneous area, regardless of the number of negative results from that area. An example of the use of the positive stop method is shown in Table 11.2, where samples PI-2 and PI-3 were not analyzed due to the positive result for sample PI-1.

If there are contradictory results from a group of samples, for example, one of nine samples of surfacing material is positive and the other eight are negative, then the inspector may want to reconsider the assumptions of the survey. Are there subtle physical differences between the positive and negative materials? Do they have a different application history? If any doubt remains about the meaning of the analytical results or if further information is needed, then the inspector should consider collecting additional bulk samples, redefining the building's homogeneous areas, or conducting additional analyses (e.g., NOB or TEM) on the samples in question.

The potential savings in analytical costs have to be weighed against the extra information obtained by analyzing all of the samples. In many cases, it may be more efficient to have all the samples analyzed at once rather than follow the positive stop method.

TABLE 11.2

Example of a Portion of a Summary Table of Analytical Results from an Asbestos Survey

Sample No.	Floor	Location	Materials/Items	Analytical Results
FT-1	3	Room 301	12′ × 12′ floor tiles	NAD
FT-2	3	Room 303		NAD
FT-3	3	Room 305		NAD
SM-1	3	Near front entrance	Popcorn ceiling texture	0.3% Chrysotile
SM-2	3	Opposite wall of Room 301		NAD
SM-3	3	Near door of Room 301		NAD
SM-4	3	Near door of Room 303		NAD
SM-5	3	Near door of janitor's closet		2.4% Chrysotile
PI-1	3	Women's bathroom, far wall	Corrugated piping insulation	39.5% Chrysotile
PI-2	3	Women's bathroom, middle		Positive stop
PI-3	3	Women's bathroom, near door		Positive stop

Note: NAD, no asbestos detected.

11.6 Hazard Assessment

11.6.1 Physical Hazard Assessment

AHERA specifies that the building inspector is to conduct a physical assessment of all friable confirmed and PACM. The management planner must then determine the hazard posed by the material and rank the hazards according to severity. The physical assessment consists of (1) assessing the condition of the material and (2) assessing the potential for future disturbance. Following the assessment, all identified ACBM are placed in one of the seven categories of condition and potential for disturbance. This subjective determination is the most difficult job of the inspector. In theory, there is no right or wrong assessment as long as the inspector can justify it. However, simply assessing a sprayed-on material as significantly damaged without explaining the nature of the damage is not acceptable under AHERA.

The inspector must view this aspect of the inspection process as if a map is being drawn for the management planner. Although most management plans are written by the same individual who performed the inspection, the assessment process—the entire inspection process—should be approached as if a total stranger will read the report. From the inspector's report, the management planner should have enough knowledge concerning the ACBM in that building to make an evaluation of the hazards posed to occupants.

Functional spaces can be used to divide the homogeneous area to provide a physical assessment on a space-by-space basis. Since the condition of a particular homogeneous material may differ in each functional space, the assessment of the homogeneous material may differ from functional space to functional space. The inspector must assess each friable homogeneous material as it appears in each functional space. The inspector does not assess the homogeneous material as a single unit, but assesses it according to its condition in each functional space.

The inspector has two options by which to assess the current condition of the ACBM. The first choice is to go from functional space to functional space and assess all ACBM in that functional space. Some functional spaces may contain several homogeneous materials. The second choice is to pick the first homogeneous material and go to each functional space in which it appears and assess the material. Either way is acceptable as long as the material is fully assessed.

11.6.2 Classifying the Condition of the Asbestos-Containing Building Material (ACBM)

The potential for building occupants to be exposed to asbestos depends on the condition of the ACBM, the likelihood of disturbance, and the potential for fibers to be transported. The latter issue is not a consideration when performing a hazard assessment under AHERA. AHERA divides the condition of friable surfacing material and TSI into three categories: good, damaged, or significantly damaged.

Good condition—Material with no visible damage or deterioration, or showing only limited damage or deterioration.

Damaged—Material that has deteriorated or sustained physical injury such that the internal structure (cohesion) of the material is inadequate or which has delaminated such that its bond to the substrate (adhesion) is inadequate, or that for any other reason lacks fiber cohesion or adhesion qualities (flaking, blistering, or crumbling); water damage, significant or repeated water strains, scrapes, gouges, mars or other signs of physical injury.

Significantly damaged—Extensive and severe damage to friable surfacing material and TSI. To decide whether the material is simply damaged or if the damage is significant in nature, the USEPA recommends the "10/25 rule" to determine if a surfacing material is damaged or significantly damaged. The 10/25 rule states that a material is significantly damaged if:

- The extent of the damage is roughly 10% of the materials and is evenly distributed throughout the material. This criteria pertains to water stains, gouges, and marks on the ACBM (see Figure 11.10).
- The extent of the damage is roughly 25% of the materials and is localized (see Figure 11.11).

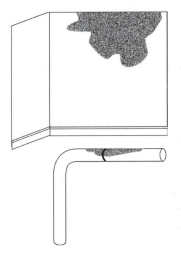

FIGURE 11.10
Examples of 10% uniform damage that result in a significantly damaged ACM classification. (Adapted from the U.S. Environmental Protection Agency.)

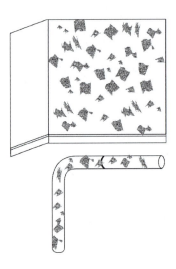

FIGURE 11.11
Examples of 25% localized damage that result in a significantly damaged ACM classification. (Adapted from the U.S. Environmental Protection Agency.)

Accumulation of powder, dust, or debris of similar appearance to the ACBM on surfaces beneath the material can be used as confirmatory evidence. Since the human eye tends to underestimate the percentage of uniform damage in a damaged material, Figure 11.10 and Figure 11.11 can be useful guides for the inspector.

The assessment of miscellaneous materials depends on the type of material to be assessed. For instance, if asbestos-containing ceiling tiles are damaged or significantly damaged, they are unrepairable, although replaceable. Floor tiles and other nonfriable miscellaneous materials, on the other hand, require no assessment if undamaged. If asbestos-containing floor tiles are damaged, such as the floor tiles shown in Figure 11.12, then it would be appropriate for the inspector to recommend removal as the response action, without assigning it a hazard category. This is true for most damaged, nonfriable ACBM.

11.6.3 Classifying the Potential for Disturbance of ACBM

Once its condition is assessed, the potential for damage or significant damage to all ACBM and PACM must also be assessed. It would be of little value to repair the damaged ACBM if the cause of the damage is not also corrected. The AHERA management planner must determine the cause of damage or potential damage and include that information in the inspection report. The following are a few types of activities that may release or resuspend fibers into the air:

- Renovation projects
- Repair and maintenance activities
- Routine cleaning

FIGURE 11.12
Significantly damaged floor tiles. (Courtesy of GZA GeoEnvironmental, Inc.)

- Operation of building systems
- Activities of occupants other than service workers
- Deterioration/aging of the ACBM

Exposure factors must also be considered when assessing the potential for disturbance. For example:

- Nonvisible materials—This refers to the common situation that if an ACBM is not immediately visible, the occupants may forget its presence and accidentally expose it in a manner that may result in fiber release.
- Accessibility—The degree to which the material can be easily contacted by a building occupant. ACBM that is easily accessible can be just as easily disturbed.
- Barriers—Are there barriers to casual contact? If so, are they permanent or temporary? How easily can these barriers be breached?
- Ventilation—Is the material in contact with the building's air stream?
- Air movement—Air erosion may occur in a return air plenum or fan room. This movement may erode the ACBM causing potential exposure problems.
- Air plenum—Is the air plenum used for return air? If so, material in this area may be dispersed throughout the building.
- Activity—Heavy activity such as in a gym or machine shop would present a greater hazard than an area such as a library, or an electrical room with limited access.
- Vibration—Vibration from mechanical equipment, subway tunnels, buses, loud noise, and other sources could eventually loosen the material.
- Character of the occupants—School occupants, and children in general, have a higher potential to disturb ACBM than the occupants of a commercial building.

Although all the exposure factors must be taken into account when assigning a potential for damage to the ACBM, the three most important factors are considered the frequency of potential contact, influence of vibration, and potential for air erosion. The difference between potential for damage and potential for significant damage plays an important role in this assessment. AHERA defines potential damage as follows:

- Friable ACBM is in an area regularly used by building occupants, including maintenance personnel, in the course of their normal activities.

- There are indications or there is a reasonable likelihood that the material or its covering will become damaged, deteriorated, or delaminated due to factors such as changes in building use, changes in operations and maintenance practices, changes in occupancy, or recurrent damage.

In addition to these conditions, the existence of major or continuing disturbance due to accessibility to building occupants, vibration, or air erosion would lead the management planner to label the ACBM as "potential for significant damage."

Changing the use of an area will change the potential for disturbance of an ACBM. Demolition or renovation activities in an area with ACBM change the potential for disturbance. In such cases, the ACBM must be removed prior to the onset of renovation or demolition activities. Other changes in usage of the functional space, such as the conversion of a maintenance room into a classroom, require reclassification of the potential for disturbance. AHERA requires periodic review of hazard assessments since unless a material is located behind a permanent enclosure, such as a wall or other airtight barrier, the potential always exists for the material to be disturbed and damaged.

11.6.4 Seven Categories of Hazards

AHERA provides seven categories by which to assess the current condition of the ACBM and the potential for damage (see Figure 11.13). The management planner uses these classifications when determining the appropriate response actions (see Chapter 12).

In most cases, ACBM with a 7 hazard ranking will require an active response action, that is, removal, encapsulation, or enclosure. ACBM with a 1 hazard ranking can most likely remain in place under an operations and maintenance (O&M) plan. The response actions to be taken for materials with hazard ranking from 2 through 6 will require the judgment of the management planner, and close coordination with the building owner, designated person, or other responsible persons in the building.

It should be noted that damaged ACBM with a low potential for disturbance is given a higher hazard ranking (4) than undamaged ACBM with a high potential for disturbance (3). This logic falls under the "one in the hand, two in the bush" concept. In other words, since damaged ACBM already has the potential to release harmful asbestos fibers into the environment—the hazard is immediate and is a priority to address. However, the management planner should not let this decision tree cloud his or her better judgment. If a library is being converted into an in-line skating rink, then the potential to disturb undamaged ACBM wall plaster is so significant that this material should take precedence over the TSI located 30 feet above the floor on a disconnected fan, even if that ACBM is already damaged.

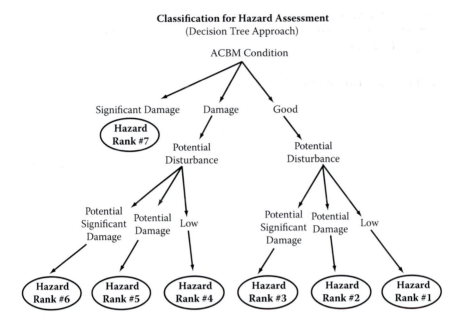

Classification for Hazard Assessment
(Decision Tree Approach)

FIGURE 11.13
AHERA hazard potential decision tree.

11.6.5 Air Monitoring for a Hazard Assessment

A hazard assessment for most airborne contaminants involves air sampling. Air monitoring for an asbestos hazard assessment is, however, not the preferred method for several reasons:

- It can be expensive.
- It only provides a snapshot of conditions at the time of the air monitoring.
- It does not allow for changed conditions.
- It typically does not measure conditions at the times of peak disturbance of the asbestos, such as when the area of interest is actively occupied by employees, or when doors open and shut, creating air movement.
- It does not take into account atmospheric changes, such as changes in temperature, pressure, and humidity that can also affect measurements.

For these (and other) reasons, EPA does not recommend—and AHERA does not mention—air monitoring for assessment purposes. However, some building owners like to use air monitoring to "spot test" for high levels of asbestos. This testing can often placate baseless fears of building occupants who discover they are working in an environment that contains asbestos. On the other hand, a high reading may indicate a problem overlooked by or invisible to the inspector.

References

Centers for Education and Training, Rutgers University of Medicine and Dentistry of New Jersey (UMDNJ). 1990. Inspecting Buildings for Asbestos-Containing Materials (AHERA Inspector).

Centers for Education and Training, Rutgers University of Medicine and Dentistry of New Jersey (UMDNJ). 1988. Procedures and Practices for Asbestos Control.

Natale, Anthony, and Levins, Hoag. 1984. *Asbestos Removal & Control: An Insider's Guide to the Business.* J. Levins Design, Inc.

12

Asbestos Abatement

12.1 Introduction

Once the asbestos hazard assessment has been completed, response actions must be determined for each asbestos-containing material (ACM) identified in each functional area. This chapter discusses the types of asbestos response actions, which collectively are known as asbestos abatement.

12.1.1 Types of Asbestos Abatement

The term "asbestos abatement" is often used synonymously with the term "asbestos removal." In fact, there are four types of asbestos abatement, the other three types being enclosure, encapsulation, and operations and maintenance. All four types of asbestos abatement are designed to remove the asbestos hazard, not necessarily the asbestos itself.

Since the most technically rigorous (and the most highly prescriptive) method of abatement is removal, most of this chapter describes asbestos removal. The other three asbestos abatement methods are discussed at the end of the chapter.

12.1.2 Regulatory Requirements for an Asbestos Abatement Project

At the federal level, the Occupational Safety and Health Administration (OSHA) and the U.S. Environmental Protection Agency (USEPA) regulate asbestos abatement projects. Their regulations have different purviews, reflecting the differing missions of the two agencies. OSHA's mission is to protect workers from workplace hazards. As such, OSHA's Asbestos Standard for Construction, codified in 29 CFR 1926.1101, is designed to protect workers who are performing the abatement operations.

USEPA's mission is to protect human health and the environment. The USEPA regulates asbestos under the Toxic Substances Control Act (TSCA), which includes the Asbestos Hazard Emergency Response Act (AHERA) and Asbestos School Hazard Abatement Reauthorization Act (ASHARA). The USEPA also regulates asbestos under NESHAP (the National Emission

Standards for Hazardous Air Pollutants; see Chapter 2). NESHAP plays a role in the establishment of work practices to minimize release of asbestos fibers during activities that will disturb ACM.

Many states have promulgated their own asbestos abatement regulations, which, by law, must be at least as stringent or more stringent than the federal laws. Certain cities, such as Philadelphia and New York City, have their own set of regulations for asbestos abatement projects. As such, the discussion presented in this chapter is for instructional purposes only and should not be construed as a summary of any particular regulation. One should consult the local regulations before planning and commencing with any asbestos abatement project.

The contractor who will be performing the abatement activities typically holds a license that was issued by the state or the city in which the project is taking place. The individual workers and the project foreman are also subject to licensure requirements within their jurisdictions.

12.1.3 Sizes of Asbestos Abatement Projects

There are three types of asbestos abatement projects, based on the volume of ACM to be disturbed:

- A large asbestos project involves the removal, encapsulation, enclosure, or disturbance of more than or equal to 160 square feet (SF) of ACM or more than or equal to 260 linear feet (LF) of ACM.
- A small asbestos project involves the removal, encapsulation, enclosure, or disturbance of between 10 and 160 SF of ACM, or between 25 and 260 LF of ACM.
- A minor asbestos project involves the removal, encapsulation, enclosure, or disturbance of less than or equal to 10 SF of ACM, or less than or equal to 25 LF of ACM.

12.1.4 Public Notifications

Prior to the beginning of the project, the contractor is required to notify the USEPA, and in some cases the state and local jurisdictions, of the beginning of the project. The notification is usually submitted in writing to one or more of the applicable regulatory agencies.

12.2 Asbestos Abatement by Removal

Asbestos abatement projects are some of the most highly proscribed activities in the sphere of environmental regulations. While the obvious number one objective of the project is to remove the designated ACM, the "first, do no harm" objective plays a critical role in asbestos removal as well. Falling under this header is the goal to prevent asbestos fibers from leaving the workplace and impacting humans outside of the workplace, and the goal of preventing detrimental health effects to the workers performing the abatement. Huge portions of the abatement project entail the installation of engineering controls to prevent the spread of asbestos fibers out of the workplace and using procedures designed to keep asbestos fibers from becoming airborne. The following sections outline some of the procedures employed in attaining these objectives.

It should be noted that although various jurisdictions have specific requirements for the preparation of the work area (and, for that matter, all other aspects of asbestos removal), what is presented here is a baseline methodology for asbestos removal projects.

12.2.1 Preremoval Preparations for All Removal Projects

12.2.1.1 Signage

On the day of mobilization, the contractor will post the appropriate notifications prominently on the building entrance and post warning signs at the entrances to the work area (see Figure 12.1). These signs are in English as well as the main language spoken by the building inhabitants.

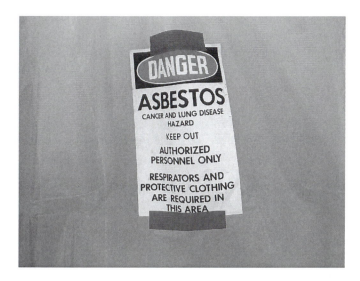

FIGURE 12.1
Asbestos warning sign. (Courtesy of M. Stuckert, Save the 905, Inc., www.RockIsland905.com.)

12.2.1.2 HVAC Shut Down

To prevent the spreading of asbestos fibers throughout the building, the building's heating, ventilation, and air-conditioning (HVAC) system should be shut down and isolated.

12.2.1.3 Preabatement Air Sampling

Air samples are collected inside and outside the work area, typically at least one day before the asbestos removal begins. The objective of the air sampling is to establish preabatement, or "background" conditions against which air quality can be judged in the course of the asbestos removal activities.

To collect these preabatement air samples, an air sampling technician sets up air pumps at various locations inside and outside the proposed work area. The air pumps are equipped plastic tubing with sample cassettes at the end of the tubing. Inside the cassette is a microfilter that is designed to allow the passage of air while retaining microscopic fibers. The pumps are set at a flow rate of 2 to 10 liters per minute and operated until at least 560 liters of air have been drawn into the cassette. The pumps are calibrated prior to pump usage, and at the beginning and the end of the project.

Once the required volume of air has been drawn by the air pump, it is shut off and the cassette is labeled and sent to a laboratory for analysis using phase contrast microscopy (PCM). In PCM analysis, all fibers greater than 5 microns in length having an aspect ratio (length to width) of at least 3 to 1 are counted. The PCM method cannot distinguish between asbestos and nonasbestos fibers, making it an inaccurate measuring tool, especially in dirty areas or areas where fibrous materials, such as clothing, are stored. Furthermore, PCM cannot detect thin fibers (less than about 0.25 micrometers in diameter), which may comprise the majority of airborne asbestos fibers in a building with ACBM.

In such cases, the more expensive transmission electron microscopy (TEM) method is employed, which allows the lab technician to distinguish asbestos fibers from nonasbestos fibers, and to observe very thin fibers.

12.2.2 Preparing the Work Area for the Large Asbestos Project

OSHA defines two classes of asbestos removal projects. Class I removal projects involve the removal of thermal system insulation (TSI) and surfacing ACM and PACM only. Projects involving the removal of miscellaneous materials, which OSHA defines as Class II removal projects, are subject to less stringent regulations (unless otherwise regulated by the local jurisdiction). The following sections describe some of the steps ordinarily used in preparing an area to undergo asbestos removal in a large, Class I asbestos removal project.

The preparation of the work area involves the construction of a containment, which, as its name implies, is an isolated work area in which the asbestos stays until it is removed in a controlled manner. The construction of a containment is a time-consuming process, involving the installation of numerous engineering controls. The various steps involved in preparing the work area are described next.

12.2.2.1 Movable Items

All movable items must be removed from the area that will become part of the containment. The objects should first be cleaned with a vacuum equipped with a high-efficiency particulate air (HEPA) filter and wet-wiped in case they are contaminated with asbestos dust. Drapes and other fabrics in the work area should be removed and either cleaned or disposed. Carpets contaminated with debris that may contain asbestos should be disposed of as asbestos waste.

12.2.2.2 Stationary Items

Stationary items, such as machinery, light fixtures, blackboards, water fountains, and toilets, that will remain within the containment should be protected from the upcoming removal activities. They first should be wet-wiped or HEPA vacuumed, and then covered with 6-mil-thick polyethylene plastic (colloquially known as "poly"), which is secured to the object with duct tape.

12.2.2.3 Construction of Decontamination Units

For large asbestos projects, once movable and stationary objects have been addressed, the next step is the construction of two decontamination units: the personal and waste decontamination units (see Figure 12.2). The decontamination units will act as controlled pathways for workers and wastes in the course of the removal activities.

The *personal decontamination unit* is designed to allow workers to move to and from the work area during abatement activities while preventing asbestos from leaving the work area. A personal "decon" (as it is typically called) consists of the following three rooms:

> *Equipment room, also called the dirty change room*—This room, which adjoins the work area, is a contaminated area where workers exiting the containment leave their equipment, footwear, hardhats, goggles, and other asbestos-contaminated apparel and protective gear. Workers, who are typically outfitted in respirators, do not remove their respirators in the equipment room, since that room is considered to be contaminated.

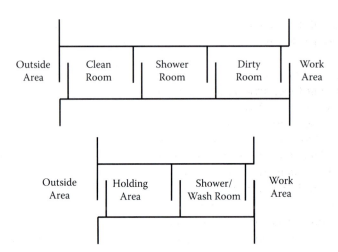

FIGURE 12.2
Personal decontamination unit (top) and a waste decontamination unit (bottom).

> *Shower room*—Workers enter the shower room from the equipment room, still wearing respirators. Here, workers use shower water to remove any residual asbestos debris that may have adhered to their bodies, hair, or respirator. The shower room is equipped with a 5-micron drain to capture wastewater before it enters the hoses that connect the shower room to the building's sanitary sewer system. The filters must be replaced on a regular basis to avoid clogging.

> *Clean room*—From the shower room, decontaminated workers enter the clean room, where their street clothing and uncontaminated materials and equipment are stored. Upon entering the clean room, the workers can dry their bodies, remove their respirators, and dress. The spent, wet respirator cartridges are disposed in the clean room. Clean rooms are typically equipped with benches, lockers for clothes and other personal effects, and nails or hooks for hanging respirators.

The personal decon is also used for worker entry into the containment. Before entering the work area, the workers use the clean room to don their respirator (equipped with new cartridges), and protective clothing, usually disposable coveralls or Tyvek™ suits. They then proceed through the shower room into the equipment room, where they retrieve the equipment they left in the equipment room and enter the work area.

The second type of decon unit is known as the *waste decontamination unit*. The purpose of the "waste decon" is for the removal of asbestos-contaminated waste from the containment. The waste decon has two rooms rather than three, since it lacks an equipment room. Asbestos-contaminated wastes, which have been placed in specially designated bags inside the containment, are carried into the first room, known as the shower/wash room, where workers, wearing full protective gear, wash residual debris from the exterior of the bags by HEPA vacuuming or wet wiping. The workers then place the bags into the adjoining holding area, where workers, who may not be wearing protective gear, take the cleaned bags and load them into a secure container or directly onto a truck for transport to a licensed disposal facility. Personal decon units and waste decon units may be combined into one unit if warranted by space limitations outside the work area.

Decon units are typically constructed with 2″ × 4″ wood frames and poly walls and floors (although plywood walls are required in some jurisdictions). Two layers of poly sheeting typically separate each of the rooms to act as *air-locks* by inhibiting the flow of air between rooms. Poly sheets are typically hung from either side of the decon, overlapping in the middle. Predesigned decon units, usually designed as trailers, are sometimes used on abatement sites, especially for outdoor work.

12.2.2.4 Critical Barriers

The next phase of construction involves isolating the work area so that the only ways into and out of the work area are through the decon units. This involves the establishment of *critical barriers* at all other openings into the work area. These openings include windows, doorways, and skylights, as well as vents and air ducts for the HVAC system, and holes in the floors, walls, and ceiling for utilities such as conduit and piping. Critical barriers are composed of two layers of 6-mil poly, which are secured to the perimeter of the opening with duct tape. The two layers are individually secured so that the inner layer of poly will remain in place if the outer layer should happen to become dislodged, thereby maintaining the integrity of the critical barrier.

If there are stray asbestos fibers on the items being plasticized, they will still be there at the end of the abatement project, thereby defeating the goal of the project to make the work area free of asbestos. Therefore, it is prudent to wipe the surfaces to be plasticized prior to covering them in plastic.

12.2.2.5 Plasticizing Floors and Walls

Once all critical barriers have been established and all stationary objects in the work area covered, the work area is *plasticized*, which involves covering the room with protective 6-mil poly. Individual poly sheets are taped to the walls and to each other with duct tape or spray adhesive (see Figure 12.3). The sheets on the floor should rise at least one foot above the floor (more is required in some jurisdictions) and covered with the poly on the walls so that they overlap, preventing creation of a seam through which asbestos fibers could escape. As with the critical barriers, each layer of poly should be installed independently, so one layer will remain in place if the other layer fails. Certain jurisdictions mandate waiting times between installation of the individual poly layers, so

FIGURE 12.3
Preparing a room for asbestos abatement. (Courtesy of M. Stuckert, Save the 905, Inc., www. RockIsland905.com.)

that it could be determined whether the first layer of poly has been adequately secured prior to the installation of the second layer of poly. Sometimes, it may be necessary to secure the poly with nails or furring strips, especially if the walls or floors are slippery or the poly is getting too heavy.

12.2.2.6 Electrical Lock-Out

Water is typically used to saturate the ACM prior to removal, creating a humid environment in the work area. To eliminate the potential for a shock hazard, the electrical supply to the work area should be de-energized and locked out before removal operations begin. This is typically done after the work area has been plasticized, so that workers can work in normal light for as long as possible. OSHA requirements regarding lock-out/tag-out should be used at all times when dealing with the de-energizing of electrical equipment. Temporary electrical service using *ground-fault interruption* (GFI) protection is brought into the work area for use in lighting the work area and for power equipment.

12.2.2.7 Establishing Negative Pressure

The final step in the preparation of the work area is the establishment of a *negative air pressure* regime. By creating negative air pressure inside the containment, air will flow from high pressure to low pressure, that is, from outside the containment into the containment. In order to escape a work area, an airborne asbestos fiber inside the containment would have to go against the air flow, which is difficult (but not impossible) to occur. The majority of airborne fibers would not be able to escape the containment, thereby protecting humans outside the containment.

The pressure gradient is maintained with the use of powered exhaust equipment. Units are placed inside the containment that are equipped with HEPA filters, which will capture airborne asbestos fibers prior to exhausting the filtered air outside of the containment (and generally outside the building).

The quantity, size, and the placement of the HEPA units, as they are typically called, are determined by the size and configuration of the work area. Figure 12.4 shows the area that was the subject of the asbestos survey described in Chapter 11. That survey identified two types of ACM: surfacing material in the hallway and piping insulation in the women's bathroom. A full containment has been constructed for the removal of the surfacing material (the removal of the piping insulation is discussed later in the chapter). Let's assume that the containment is 15′ × 60′ in area, the ceiling is 8′ high, and regulations require that there be at least one air change every 15 minutes in the work area. The volume of air to be changed every hour in the work area is calculated using the following formula:

Volume of air to be moved = Volume of room × Number changes per hour

which in this example is calculated to be:

$$8' \times 15' \times 60' = 7200 \text{ ft}^3 \times 4 \text{ changes per hour} = 28800 \text{ ft}^3/\text{hr}$$

A typical commercially available HEPA unit may have an air handling capacity of 1000 to 2000 ft^3 per minute, which is equivalent to 60000 to 120000 ft^3/hr. Therefore, for this space, one HEPA unit operating at 1000 ft^3/min will provide more than enough air movement to meet regulatory requirements. Note that the number of HEPA units is not based on the quantity, type, or location of the ACM itself.

HEPA units should be placed to avoid "dead" spaces, or places where no air is flowing. It should also be placed far from the "make-up" air, which is the clean air flowing into the containment from the decon units, to promote efficient air flow. The HEPA unit shown on Figure 12.5 is situated at the opposite end of the containment from the personal decon for that reason. The HEPA unit vent, which is a flexible, plastic duct, extends to a window located in a nearby room through which the filtered air is vented.

Once the HEPA units are set up and operating, the establishment of the negative air pressure regime should be verified. For instance, the poly

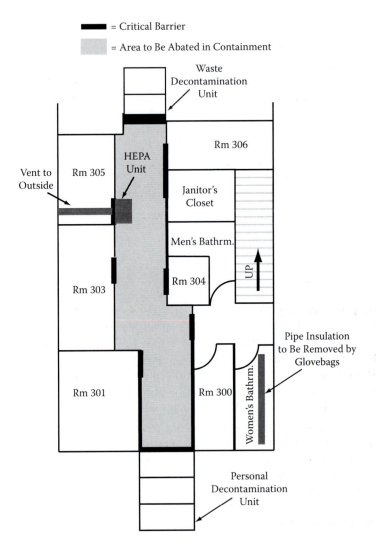

FIGURE 12.4
Layout of a building area about to undergo asbestos removal.

sheeting separating the rooms of the decon unit should be pointed into the work area. Smoke tests may be performed to verify that air from outside the decon unit is being drawn into the work area. For quantitative measurements of differential air pressure, a manometer or a magnahelic gauge often is attached to one of the decon units or other locations where such data is desirable. A pressure drop of 0.02 to 0.03 inches of water across the barrier is usually indicative of an adequate negative air pressure regime. Higher pressure differentials may create too much air turbulence for workers to adequately perform their duties in the work area.

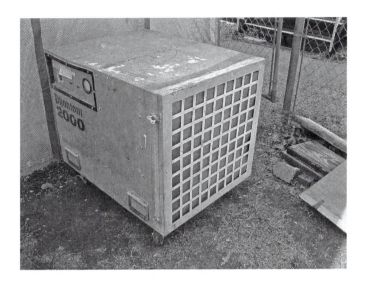

FIGURE 12.5
HEPA filtration unit outside of a containment. (Courtesy of M. Stuckert, Save the 905, Inc., www.RockIsland905.com.)

12.3 Removing the Asbestos

The containment is now complete and a negative air pressure environment has been established. It is time to remove the asbestos. Workers enter the clean room of the personal decon unit, suit up in protective clothing, don their respirators and confirm the fit, gather their equipment, and enter the containment. Typically they sign a sign-in sheet that is posted near the entrance of the personal decon unit.

12.3.1 Wetting the Material

To minimize the creation of airborne asbestos fibers, ACM is wetted with amended water (see Chapter 11) prior to removal. Dry removal is allowed only in special circumstances, such as when there is an electrical source in the work area that cannot be de-energized, creating a danger of electrocution. Wetting the material is accomplished using an *airless sprayer*, which operates at a low pressure, thus avoiding the haphazard dislodging of the ACM from the substrate. High-pressure wetting, as through a garden hose, may dislodge the ACM before it is adequately wet, resulting in an unacceptable release of asbestos fibers into the workplace.

Once wetted, the material may absorb the water resulting in a drying of the surface of the material in the course of the removal operations. Therefore, it is important to keep the material "continually" wet.

12.3.2 Two-Stage ACM Removal

ACM removal is accomplished in two stages. First, there is the *gross removal*, generally performed using scraping tools for surfacing material and many miscellaneous materials, or a utility knife or other cutting device for thermal system insulation (see Figure 12.6). Best management practices dictate catching the removed material before it falls to the ground, and picking up fallen material before it dries up and emits asbestos fibers into the workplace, then placing the removed ACM in properly labeled waste bags, taking care not to tear the bags.

The second removal stage, known as *fine cleaning*, involves removing residual ACM that has adhered to the substrate. This is typically accomplished by a combination of fine brushing and wet wiping. HEPA vacuums may also be employed to remove ACM in relatively inaccessible locations in the work area.

Once completed, the abated surfaces are inspected for the presence of residual ACM or dust that may contain asbestos fibers. Fine cleaning should continue until no visible ACM or dust is visible on the abated surfaces.

12.3.3 Glovebag Removal

An often-used, specialized procedure for the gross removal of ACM involves the use of a *glovebag*. The glovebag is a plastic bag designed with two sleeve gloves, a tool pouch, and a small opening that may be used for the insertion of airless sprayers or HEPA vacuum nozzles. The glovebag is designed to act as its own containment for the safe removal of thermal system insulation

FIGURE 12.6
Workers removing asbestos inside a containment. (Courtesy of M. Stuckert, Save the 905, Inc., www.RockIsland905.com.)

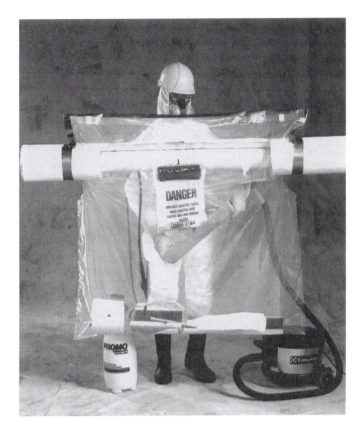

FIGURE 12.7
A glovebag attached to a horizontal pipe. (Courtesy of Grayling Industries.)

from horizontal linear features such as piping, as shown in Figure 12.7. Using glovebags where appropriate can result in huge savings in time and costs, since it avoids construction of a full containment in the work area.

To fit the glovebag onto the pipe, slits are cut in its sides, and the glovebag is draped over the ends of the pipe. Tools that will be used in removing the ACM from the pipe are placed in the tool pouch, and the slits and the top of the glovebag are then taped shut, thus sealing the glovebag to the pipe.

Prior to the start of the glovebag removal procedure, its integrity should be tested by inserting a smoke tube through the opening in the bag and observe whether any smoke leaks out of the bag. Workers will apply duct tape to any openings in the bag that allow smoke to escape.

Once it is determined that the glovebag is airtight, the worker inserts the nozzle of an airless sprayer into the opening and wets the material to be removed. The worker can then reach into the bag through the sleeve gloves, access whatever tools are in the tool pouch, and safely remove the ACM, which falls to the bottom of the bag. Damp rags earlier placed inside the glovebag can then be used for the fine cleaning process. Once completed, the tools are returned to the tool pouch, which can then be pulled through the glovebag, tied off, and removed intact, eventually to be decontaminated in the personal decon unit along with the worker. A HEPA vacuum is then used to remove air from the spent glovebag, which is then twisted and taped off and cut from the pipe. The entire glovebag, with the ACM waste at the bottom, can then be safely placed into an asbestos waste bag for eventual disposal.

The piping insulation located in the women's bathroom on Figure 12.4 is to be removed using glovebags. No personal or waste decon units need to be next to this area, since no containment needs to be constructed. Once the insulation has been removed, the glovebags will be removed from the pipe, tied off, and disposed through the waste decon while the workers exit the work area through the personal decon.

12.3.4 Waste Removal

Asbestos-containing wastes are placed in specially marked plastic bags for eventual disposal at a licensed landfill (see Figure 12.8). These wastes include materials removed during the preparation of the work area, such as HVAC filters, and possibly drapes and carpeting that may have been contaminated with asbestos fibers. They also include shower filters from the personal decon unit.

FIGURE 12.8
Bagged asbestos waste inside a waste decon. (Courtesy of M. Stuckert, Save the 905, Inc., www.RockIsland905.com.)

Asbestos waste bags have warning labels similar to those used on signage for the asbestos project workplace. They also are labeled with the regulatory classification of their waste type. Asbestos wastes are double-bagged so that there is no release of asbestos fibers if the outer bag is damaged. The bags are stored in the holding area of the waste decon prior to being loaded onto appropriately licensed and labeled trucks for transport to a landfill that is licensed to accept asbestos waste.

12.3.5 Lockdown Encapsulation

Lockdown encapsulation is the application of a protective coating or sealant to a surface from which ACM has been removed. Though the substrate may appear to be clean, microscopic asbestos fibers may be present, or small amounts of ACM may still remain in cracks or crevices that were inaccessible during the removal of the ACM. Lockdown is designed to prevent these fibers from becoming airborne.

The lockdown material is applied once the surface has been fine cleaned and allowed to dry. The material is a liquid, similar to paint, that, once dry, forms a membrane over the surface (also known as a *bridging encapsulant*) or can penetrate into the material and bind its components together (a *penetrating encapsulant*).

12.3.6 Two-Stage Cleaning and Poly Removal

Once the encapsulant has dried, the top layer of poly is thoroughly inspected for the presence of asbestos debris or dust. If debris or dust is noted, the area is recleaned and reinspected. Once the work area passes inspection, the first layer of sheeting can be removed. The removed poly is double-bagged in asbestos waste bags for removal from the work area.

The process is then repeated for the remaining layer of poly. Once this layer passes the visual inspection, it too is removed using the procedure described earlier. Some jurisdictions mandate a waiting period between these two rounds of cleaning to allow any airborne asbestos fibers to settle or to get filtered from the air by the HEPA units. With only the critical barriers and decon units remaining, a third round of cleaning is then performed, again in some jurisdictions after a mandated waiting period. The work area is now ready to be tested for clearance (see Section 12.4).

12.3.7 Abatement Air Monitoring

In the course of the asbestos removal and cleaning activities, three types of air samples are collected:

- Perimeter air samples
- Area air samples
- Personal air samples

Area air samples are collected inside the containment. They are designed to determine the concentrations of airborne asbestos fibers, which are used to evaluate the effectiveness of the contractor's efforts to minimize the amount of asbestos fibers released into the air, and whether the workers are using the proper level of respiratory protection.

Perimeter air samples are collected around the outside of the containment where asbestos fibers are most likely to escape. The purpose of perimeter air monitoring is to assess the effectiveness of the engineering controls to isolate the work area and prevent asbestos fibers from leaving the work area. On Figure 12.4, perimeter air sampling locations may include the outside of the two decon units, outside the building near the HEPA unit exhaust, and near the critical barrier that was set up to isolate the staircase (in the foyer between Rooms 300 and 304).

Together, area air samples and perimeter air samples are known as *daily air samples*, because they are collected every day that the asbestos removal activities are conducted in the work area. Five inside air samples and five outside air samples must be collected each day for a large asbestos abatement project (which means that one other location would need to be selected for the collection of a perimeter air sample in Figure 12.4).

The air samples usually are analyzed using PCM methods (or TEM methods if warranted). The results are compared to the regulatory standard of 0.01 fibers per cubic centimeter (f/cc) as well as the results from the preabatement air monitoring. If the air monitoring results suggest a breach in the work area containment, all work is stopped until the source of the problem is identified and fixed.

Personal air samples are specifically designed to evaluate worker exposure to airborne asbestos fibers and are required by OSHA. These measurements can be used to calculate the volume of asbestos fibers a worker is inhaling and determine whether the respirator being worn is adequately protecting the worker.

A personal air monitoring pump is strapped to the worker's waist and the plastic tube is taped to the shoulder so that the cassette is near their face, thus providing a good approximation for the air that the worker is actually breathing. The worker then conducts his or her daily activities as usual. At the end of the day, the pump is turned off and the cassette is sent to a laboratory for PCM analysis. OSHA regulations require monitoring of each job function inside the workplace. For instance, if workers are loading out waste bags while gross and fine removal are being performed, then at least one worker performing the load-out, one worker performing the gross removal, and one worker performing the fine removal need to be monitored. In all, OSHA requires monitoring of at least 25% of the total number of workers in the work area. Personal air monitoring is not required if workers are already outfitted in the maximum protective respiratory protection (which is a full-face, supplied air respirator operating in pressure demand mode), or if other daily results indicate that workers are being adequately protected and the work being performed inside the work area has not changed. Workers must be informed of the results of the personal air monitoring analyses.

12.4 Clearance Air Monitoring

At the conclusion of the removal and cleaning activities, *clearance air monitoring* is performed to evaluate whether a work area is ready for reoccupancy. In clearance air sampling, worst-case conditions are simulated by using *aggressive sampling techniques*, the assumption being that if airborne asbestos fiber concentrations are acceptable under this condition, fiber concentrations would be acceptable under any other condition.

Prior to the collection of clearance air samples, a leaf blower or a floor fan is activated inside the work area, pointed toward the abated surfaces, to stir up any remaining dust and debris that may contain asbestos. If the resulting analyses meet regulatory standards of 0.01 f/cc, then the area is ready for reoccupancy. Although PCM analyses are often used for the analysis of the clearance air samples, some jurisdictions require the TEM analysis for this most important of air sample analyses. AHERA requires TEM analysis for clearance air samples in public schools.

For a large asbestos project, five air pumps are set up inside the containment and fitted with cassettes. The pumps are operated at high flow rates so that they process at least 1800 liters of air for PCM analysis, or 1250 liters of air for TEM analysis, are processed.

12.5 Health and Safety Considerations

Throughout the asbestos project, the health and safety of the asbestos workers is of paramount concern. Asbestos fibers are hardly the only hazards that concern workers in the work area. There are numerous other hazards that are present during asbestos removal operations that can result in an injury to a worker. Most of the hazards are those typically associated with construction projects, such as:

Electrocution—Especially given the presence of water around electrical equipment in the work area. As described earlier in the chapter, GFIs are mandatory for temporary lighting on asbestos removal projects.

Ladders and scaffolding—There are specific OSHA requirements for the design and usage of ladders and scaffolding.

Slips, trips, and falls—Walking around on damp plastic sheeting in disposal footwear can be hazardous for workers. In addition, the limited visibility afforded by the wearing of respirators can result in
workers bumping into equipment. Tools carelessly laying around
the work area can also create trip hazards for workers.

Fire—The use of flammable materials and ignition sources should be
minimized in the work area. Emergency exit procedures should be
established and clearly marked in the case of a fire in the work area.

Heat stroke—Workers typically stay inside the containment for several
hours in protective clothing that may not allow for adequate air
exchange with their skin. In addition, since they are wearing respirators, they cannot drink liquids in containment. Therefore, heat exhaus-
tion and heat stroke are real concerns to workers in the work area.

12.6 Small and Minor Asbestos Projects

Small and minor asbestos projects require less preparation and air monitoring than large asbestos projects, although the need for thorough cleaning
and attention to worker health and safety does not change. For a small asbestos project, a waste decon unit is not required. No decon units are required
for a minor asbestos project. In fact, isolation of a minor asbestos project
in most jurisdictions need not be more than placing plastic sheeting on the
ground, posting signs in the appropriate places, and running yellow caution
tape across the entrance to the work space. The containment is usually no
more than a tent constructed of poly sheeting taped across the surfaces in the
work area that require protection.

No area or clearance air monitoring is required for minor asbestos projects. No air monitoring is required in the course of a small asbestos projects. Background, preabatement, and clearance air monitoring is required for
small asbestos projects, but only three air samples are required inside and
outside of the work area.

12.7 Abatement by Encapsulation or Enclosure

Abatement by encapsulation or enclosure is much less time-consuming than
abatement by removal and therefore less expensive. These abatement methods are by nature temporary, since the ACM is still present at the end of
the abatement project: they represent pathway removal rather than source

removal. These methods are desirable in certain situations, especially where the cost of removal is a deterrent or where removal would result in an unacceptable shutdown of part or all of an operating facility.

For both encapsulation and enclosure, the work area is first equipped with warning signs and established with the use of critical barriers as described earlier in this chapter. Only personal and clearance air sampling are conducted. Areas that may be disturbed in the course of the abatement activities are sprayed with amended water and kept damp to reduce airborne asbestos fibers. Loose or hanging ACM should be removed prior to encapsulation or enclosure using minor asbestos project procedures. The work area is then ready for abatement.

Abatement by encapsulation entails treating the ACM with a penetrating encapsulant that penetrates into the ACM and binds it tightly to the substrate. The encapsulant should be applied so that the entire ACM is adequately covered. Once the encapsulant has dried, clearance air samples are collected using aggressive sampling techniques. The critical barriers can be removed and the area reoccupied once acceptable clearance air monitoring results have been obtained.

Abatement by enclosure entails surrounding the ACM with an airtight enclosure that will not allow the emission of asbestos fibers into the breathable air. Such enclosures could be as small as a metal or hard plastic sheathing around piping insulation or as large as the construction of an airtight wood or metal wall around an entire building. The latter structure is sometimes applied to a building that contains large amounts of asbestos, such as an old building at a power plant, where either a temporary solution to the asbestos hazard is desirable for cost considerations or the building is unsafe to enter. In all cases, the workplace is prepared for abatement as described earlier, and the enclosure is constructed using safe work practices. Clearance air samples are collected once the enclosure has been constructed, and the work area is reoccupied once acceptable clearance air monitoring results have been obtained.

12.8 Operations and Maintenance for In-Place Asbestos-Containing Material (ACM)

Operations and maintenance (O&M) activities involve managing ACM in place. This includes periodic inspections by properly trained personnel, and the repair or small-scale removal of damaged ACM. Most O&M activities can be considered minor asbestos projects and are subject to the aforementioned procedures.

Unlike hazardous substances and petroleum products, asbestos will not go away with time. Therefore, O&M is necessarily a temporary method of abatement, since ACM will eventually degrade and require repair or removal.

AHERA requires each local education agency (LEA) to have and implement an O&M plan to prevent exposing children to an asbestos hazard. Only ACM are covered in the O&M plan, although it is a best management practice to include all ACM, including nonbuilding materials. The designated person is in charge of the implementation of the O&M plan, which includes scaled drawings showing the locations of the ACMs, and information regarding its asbestos content and hazard assessment. The O&M plan also includes documentation regarding notifications to workers, tenants, and building occupants; training and personal protection requirements for workers who handle ACM; training records for workers; and information on the history of ACM repair and removal. The O&M plan is necessarily a "living document," since the status of the ACM is bound to change over time.

References

Centers for Education and Training, Rutgers University of Medicine and Dentistry of New Jersey (UMDNJ). 1988. Procedures and Practices for Asbestos Control.

New York State Department of Labor. 2004. Industrial Code Rule 56 (12 NYCRR 56).

U.S. Environmental Protection Agency, 40 CFR 763–Asbestos.

U.S. Environmental Protection Agency. June 1985. Guidance for Controlling Asbestos-Containing Materials in Buildings. EPA 560/5-85-024. Office of Pesticides and Toxic Substances.

U.S. Environmental Protection Agency. July 1990. Managing Asbestos in Place: A Building Owner's Guide to Operations and Maintenance Programs for Asbestos-Containing Materials. 20T-2003. Office of Pesticides and Toxic Substances.

U.S. Environmental Protection Agency. January 1996. How to Manage Asbestos in School Buildings. EPA-910-B-96-001. Office of Waste and Chemical Management.

13

Lead-Based Paint Surveying and Abatement

13.1 Introduction

13.1.1 Lead Hazards

Lead is perhaps the most pervasive of all of the environmental hazards. It is a hazard in the air, in drinking water (see Chapter 14), in rivers and lakes, in soils, and, as discussed in this chapter, in paint. The reason that lead is so pervasive is because, like asbestos, it has qualities that made it a desirable additive in construction materials, fuels, and a host of other products. Because it is so widespread, it is regulated under almost every major federal environmental statute, including:

- Comprehensive Environmental Response, Compensation, and Liability Act (CERCLA; Superfund)
- Resource Conservation and Recovery Act
- Toxic Substances Control Act
- Safe Drinking Water Act
- Clean Water Act
- Clean Air Act

The federal Occupational Safety and Health Administration (OSHA) even has its own Lead Standard for Construction (29 CFR 1910.62) to protect workers from lead in the workplace.

13.1.2 History of Lead-Based Paint

Lead had been a component of paint possibly as long as paint itself has existed. Among the reasons for its widespread usage are its abundance in the earth's crust, malleability, and its durability. Unlike asbestos, which was first used on roofs in the mid-1800s and not inside buildings until the late 1800s/early 1900s, there is no building in the world that is too old to have paint that contains lead (see Figure 13.1).

FIGURE 13.1
A can of lead-containing paint.

By the 1920s, nearly all European countries had banned leaded paint. However, it was common in the United States until industry voluntarily lowered the allowable lead concentration in paint in 1960, although lead content in paint had been decreasing since the 1940s due to the health concerns it posed to residents. Paint in homes constructed before 1960 almost always contains lead.

The first official ban of lead-based paint in the United States was in 1977 when the Consumer Products Safety Commission (CPSC) banned the use of lead in paint in residences. However, the ban did not pertain to nonresidential buildings, such as offices and factories. In 1992, the U.S. Congress passed the Residential Lead-Based Paint Hazard Reduction Act, which set the limit of lead in paint at 0.5% by weight, or paint that contains lead at a concentration greater than 1.0 milligrams per square centimeter (mg/cm^2), hereafter known as *lead-based paint* (LBP). Table 13.1 shows the frequency of occurrence of LBP inside and outside a typical dwelling in the United States based on the age of the building.

13.2 OSHA Lead Standard

Most of the laws and guidelines regarding lead in paint are targeted toward residences and places where children under the age of 6 or 7 are commonly present (for instance, community centers and day care centers). Conversely, workplace lead hazards are regulated by federal OSHA and state equivalents,

TABLE 13.1

Percentage of All Paint That Is Lead-Based, by Year and Component Type

Year	Interior	Exterior
Walls/Ceiling/Floor		
1960 to 1979	5%	28%
1940 to 1959	15%	45%
Before 1940	11%	80%
Metal Components[1]		
1960 to 1979	2%	4%
1940 to 1959	6%	8%
Before 1940	3%	13%
Nonmetal Components[2]		
1960 to 1979	4%	15%
1940 to 1959	9%	39%
Before 1940	47%	78%
Shelves/Others[3]		
1960 to 1979	0%	
1940 to 1959	7%	
Before 1940	68%	
Porches/Others[4]		
1960 to 1979		2%
1940 to 1959		19%
Before 1940		13%

Source: U.S. Department of Housing and Urban Development (HUD), 1997, Guidelines for the Evaluation and Control of Lead-Based Paint Hazards in Housing.

Note: The information provided in this table is based on a survey conducted in 1990. These data are from a limited national survey and may not reflect the presence of lead in paint in a given dwelling or jurisdiction.

[1] Includes metal trim, window sills, molding, air/heat vents, radiators, soffit and fascia, columns, and railings.
[2] Includes nonmetal trim, window sills, molding, doors, air/heat vents, soffit and fascia, columns, and railings.
[3] Includes shelves, cabinets, fireplace, and closets of both metal and nonmetal.
[4] Includes porches, balconies, and stairs of both metal and nonmetal.

where they exist. Therefore, all nonresidential lead assessments are guided by the OSHA lead standard, which is discussed in this section.

Since it is OSHA's mission to protect workers from workplace hazard, the OSHA lead standard addresses worker exposure to lead. The mere presence of lead triggers the OSHA standard; there is no minimum threshold as there is in residential buildings.

Determining whether lead is present in paint in the workplace is done by *paint chip analysis*. In this method, a piece of paint is physically carved off its building component (known as the *substrate*) and analyzed for the presence of lead. The lead analysis may be performed on the premises by a mobile laboratory or sent to an off-site laboratory for analysis. Typically, off-site analysis is able to provide a higher degree of quality control, although more immediate results will be obtained from using a readily available mobile laboratory. OSHA does not recognize the more commonly used x-ray fluorescence method (which is described in Section 13.3.4) as a method to assess the presence of lead paint in the workplace.

The main pathway for lead into the workers' bodies is by inhalation, either through inhalation of lead-containing dust or inhalation of lead-containing vapors. OSHA assumes that all lead on surfaces can become airborne by the generation of either dust or vapors. The OSHA *permissible exposure limit* (PEL) for lead is based on what is known as a *time-weighted average* (TWA).

Figure 13.2 is a graph showing hourly airborne lead readings collected over an 8-hour period (from 8 am to 4 pm). The readings range from 0.0 to 8.0 micrograms per cubic meter ($\mu g/m^3$), with an average of the nine readings being 3.0. This is the TWA of the readings over an 8-hour period. The OSHA PEL for lead in the workplace is 50 $\mu g/m^3$ of air, calculated as an 8-hour TWA, and the action level at which worker monitoring for lead in air is required is 30 $\mu g/m^3$. Some compounds have acute toxicity limits as well as PELs; lead does not. More information on OSHA PELs and acute exposure limits for

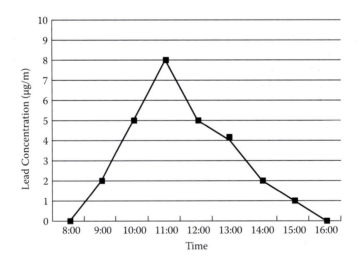

FIGURE 13.2
Hourly measurements used to estimate a time-weighted average.

various compounds and elements is provided in the guidebook published by the *National Institute for Occupational Safety and Health* (NIOSH), a division of the federal Centers for Disease Control and Prevention (CDC). The NIOSH guide is available online at www.cdc.gov/niosh/npg/default.html.

The three principal categories of work-related exposures to lead are demolition activities, renovation activities, and lead removal activities, the latter of which is discussed later in this chapter. If lead is to be disturbed by worker activity, OSHA requires the performance of an *exposure assessment* to assess potential lead hazards to workers. To conduct an exposure assessment, OSHA requires the collection of personal air samples representative of a full shift, including at least one sample for each job classification in each work area either for each shift or for the shift with the highest exposure level. For instance, if renovation activities occurring in an area where lead is known to be present include the demolition of cinder block walls, arc welding to cut steel beams, and the running of electrical conduit, then the employer must conduct three separate tests to assess the lead hazard in the workplace. If the employer knows from previous experience or from particular knowledge of the specific workplace that one activity will entail a higher exposure to lead than the others, then the employer can run tests just on that one activity.

Until the exposure assessment is performed, the employer is required to assume that the worker is being exposed to lead at a concentration above the PEL and is required to take the necessary precautions to protect the worker from that exposure. Those precautions may include removal of the lead from the workplace prior to the onset of renovation activities, placing the worker in appropriate respiratory protection, or limiting the worker's exposure by decreasing the amount of time the worker is conducting activities that disturb the lead.

13.3 U.S. Department of Housing and Urban Development Guidelines for Lead-Based Paint Inspections

The rest of this chapter is a summary of the very useful and comprehensive "Guidelines for the Evaluation and Control of Lead-Based Paint Hazards in Housing," published by the U.S. Department of Housing and Urban Development (HUD). Although it is not a regulation, it is the *de facto* standard for lead outside OSHA's jurisdiction. This discussion is not meant to be a substitute for that guide, and is certainly not meant to enable the reader to go out and perform any of the activities described here. Aside from this book being a survey guide and not a training manual, many of the activities described in this chapter require special licensing from either the U.S. Environmental Protection Agency (USEPA) or the various state governments, and all of the activities require direct field experience under the tutelage of a licensed practitioner to ensure the proper performance of the activity.

13.3.1 Lead-Based Paint and Lead Hazards

By itself, lead in paint does not present a health hazard since it does not become dust or sublimate into gaseous form readily at room temperatures. However, when allowed to chip, blister, or delaminate, as shown in Figure 13.3, it becomes a hazard via the ingestion pathway to young children, whose neurological systems are rapidly developing and who tend to put everything in their mouths, especially toddlers and other teething youngsters. In addition, due to their lesser body weight and faster metabolisms, young children have a much lower threshold for lead intake than do older children or adults. Under HUD guidelines, a lead hazard is present when at least one of the following conditions exist in a residence:

- Levels of lead above HUD standards in dust
- Contains lead at a concentration greater than 40 micrograms per square foot ($\mu g/ft^2$) on floors
- Contains lead at a concentration greater than 250 $\mu g/ft^2$ on interior window sills or 400 $\mu g/m^3$ in a window well
- Damaged or deteriorated LBP surface due to contact with another building component
- Chewable LBP surface with teeth marks
- Any other damaged or deteriorated LBP surface

FIGURE 13.3
Peeling paint on the outside of a window. (Courtesy of GZA GeoEnvironmental, Inc.)

There is another factor that makes young children more vulnerable to lead hazards. As most young parents will attest, toddlers tend to sink their newly emerged teeth into anything and everything. A window sill can provide a tempting teething surface to a toddler. If the window sill has LBP, and there is evidence of teeth marks on its surface, then a lead hazard exists. Colored paint chips that have delaminated from the walls or ceilings pose an attractive item for a young child to taste or eat.

Although lead can become airborne, it does so only in dusty conditions unlikely to be encountered in a residential setting, and it does not easily enter the gaseous phase. Therefore, the ingestion pathway is generally the only pathway considered in defining a lead hazard.

Surface soils can also be sources of lead to young children, since young children commonly will touch dirt and then put their dirty hands in their mouths, creating a direct ingestion pathway into their bodies. Soils can become contaminated from the delamination of LBP from exterior surfaces or the creation of lead-containing dust by the abrasion of LBP on exterior surfaces. Lead can also originate from atmospheric, anthropogenic sources. Any exposed play area with surface soils (i.e., soils within 6 inches of the surface) that contain more than 400 milligrams per kilogram (mg/kg; also called parts per million [ppm]) of lead are considered hazardous to young children.

13.3.2 Lead Paint Risk Assessment

To determine if a lead hazard exists in a building where young children are present, a *lead paint risk assessment* is performed. Chapter 5 of the HUD guidelines describes in full detail the procedures to use when conducting a lead paint risk assessment and is summarized here with some modifications and simplifications.

Regulations require that a USEPA-licensed risk assessor perform lead paint risk assessments. Lead paint risk assessments are not required in post-1978 buildings or buildings where all of the lead-based paint has been removed. It is permissible to perform a lead hazard screen for dwellings that are in good condition to determine if a full risk assessment is warranted. However, if the building is in poor condition or was constructed before 1960, then a full lead paint risk assessment is warranted. Without a lead paint risk assessment, HUD requires that one assume that all painted surfaces are LBP in applicable housing, and that lead from paint is present in dust inside and outside the building and in the soils surrounding the building.

Performing a lead paint risk assessment entails the following steps:

- Research
- Site walk-through
- Design the sampling program

- Sampling (dust, soil, and/or paint)
- Evaluation of analytical results
- Control identified lead hazards

13.3.2.1 Research

As with most environmental investigations, the first step is research, which entails gathering all available information about the subject matter prior to designing and implementing a sampling program. The risk assessor should interview the building owner and building manager regarding the history of LBP in the building. Has there been prior testing of LBP? Removal of old paint? Post-1978 renovations or additions to the building? Where are the children's play areas? Have the exposed soils in the play areas been tested for the presence of lead?

Visiting the local health office and interviewing the local health official often yields valuable information in the research phase of the project. Testing of lead levels in blood samples collected from children may be available, as would records of complaints from residents and parents.

The risk assessor should not blindly accept the conclusions of the documentation, but should evaluate it against the criteria for lead inspections and removals. The risk assessor should also integrate the documentation with information gathered at the interviews and his/her own personal visual assessment to assess the veracity of the existing documentation.

13.3.2.2 Site Walk-Through

Once the research portion of the lead paint risk assessment is completed, the risk assessor should conduct a walk-through of the building (or portions of the building frequented by young children). In a visual assessment, the risk assessor looks for deteriorating paint surfaces; areas of visible dust accumulation; painted surfaces that are "impact points" or are subject to friction (both of which will eventually lead to paint damage and possible lead release); and painted surfaces that a child may have chewed. Any of these observations would indicate the possible existence of a lead hazard. It should be noted that if the potential paint hazard appears to have been caused by water intrusion, then the water intrusion must be addressed before addressing the lead hazard, since failure to address the source of the lead hazard will cause the lead hazard to return after it has been mitigated.

An inspector (not necessarily the risk assessor) will observe the painted surfaces, and classify them as good or poor. Table 13.2 (Table 5.3 in the HUD guidelines) provides a template by which to assign rankings of the condition of the paint. As noted in the table, the ranking is affected by either the quantity or the percentage of deteriorated paint on a given component.

TABLE 13.2

Categories of Paint Film Quality

Type of Building Component	Total Area of Deteriorated Paint on Each Component		
	Intact	Fair	Poor
Exterior components with large surface areas	Entire surface is in good condition	≤10 square feet	>10 square feet
Interior components with large surface areas (walls, ceilings, floors, doors)	Entire surface is in good condition	≤2 square feet	>2 square feet
Interior and exterior components with small surface areas (window sills, baseboards, soffits, trim)	Entire surface is in good condition	≤10% of the total surface area of the component	>10% of the total surface area of the component

13.3.2.3 Design and Implementing a Sampling Program

Once the inspector has performed the inspection, the risk assessor (they can be the same person) will assess the potential hazards by designing a sampling plan to quantify the presence of lead inside and outside the building. Three types of testing will be conducted: dust sampling, soil sampling, and paint sampling.

Dust sampling would be conducted if dust was identified in an area that could contain LBP or at one time could have contained LBP (if the building is in a deteriorated condition). Dust sampling should be biased toward where young children are likely to be present, for instance, at the entryway of the building, in the child's principal play area (usually the TV room, living room, or dining room), the child's bedroom, the kitchen, and the bathroom. In multifamily buildings, the risk assessor should collect dust samples from common areas, such as staircases and laundry rooms. Paint on furniture, while definitely a potential hazard to the young child, is generally considered to be the responsibility of the owner of the furniture and is beyond the scope of a standard lead paint risk assessment.

Wipe sampling is the preferred method of dust sampling (see Figure 13.4). Manufacturers of wipes specify the sampling method to be used. Typically, they involve placing a treated piece of gauze of per-determined size on the surface to be tested. The wipes are then placed in a clean package and sent to a laboratory for analysis. Wipe sampling should include the collection of wipe "blanks," which are lead-free wipes, and "spiked samples," which are wipes that are spiked with lead, to test the accuracy of the laboratory's analyses.

Usually each wipe sample is collected and analyzed individually, although multiple wipe samples can be composited and analyzed as one sample. Compositing is often done to save money, but composite samples often are

FIGURE 13.4
Wipe sampling of a painted surface.

only useful as a negative screen. In other words, if three wipe samples are composited, and the result of the analysis is less than 1/3 the regulatory limit, then all three areas tested must be below the regulatory limit for lead. If the analysis shows lead above 1/3 the regulatory limit, then it is possible that one of the individual wipe samples failed the analysis, in which case the composite sample result is inconclusive for all three surfaces tested. Unless the inspector is willing to classify all of the tested surfaces as LBP, all three individual wipe samples would need to be recollected and reanalyzed, since there would be no way of knowing which individual wipe sample caused the failure. Composite samples must be from the same dwelling and from the same building component (floor, window sills, etc.). No more than four wipe samples are allowed to be composited under HUD guidelines.

When collecting soil samples for lead content, the same concepts used in dust sampling apply. Sampling should be biased to areas where young children are most likely to play, such as outdoor play areas, sandboxes, and vegetable gardens. At least two soil samples must be collected—one from the principal play area, and one from bare soil areas in front or behind the dwelling. Samples should be collected from the top of the soil, since the child will most likely contact that soil. Composite soil sampling can be performed, although it carries the same limitations as described earlier for dust sampling. Soil sampling procedures are discussed in Chapter 6.

Drinking water sampling is not required under a lead risk assessment, although it is advisable in order to obtain a full picture of the potential lead exposure in a dwelling (see Chapter 14).

TABLE 13.3

Minimum Number of Targeted Dwellings to Be Sampled among
Similar Dwellings

Number of Similar Dwellings	Number of Dwellings to Sample
1 to 4	All
5 to 20	4 units or 50% (whichever is greater)
21 to 75	10 units or 20% (whichever is greater)
76 to 125	17
126 to 175	19
176 to 225	20
226 to 300	21
301 to 400	22
401 to 500	23
501+	24 + 1 dwelling for each additional increment of 100 dwellings or less

Note: Random sampling may require additional costs.

When paint and dust testing are warranted in a multifamily dwelling (defined as a dwelling with more than four units), it is unreasonable to expect the risk assessor to obtain access to all of the units. Table 13.3 (which is Table 5.6 in the HUD guidance document) provides HUD's minimum requirements for the number of dwellings to sample in a multi-family dwelling. Reasonable assumptions would then be made about the other units based on the results of the units sampled. The designated number of units should be randomly selected, except where lead hazards are known to exist or have existed based on prior testing, there is knowledge of elevated lead in blood levels of children, lead violations have been noted by a public health inspector, and so forth. Sampling should be biased to units in which such lead hazards are known to be present or have been present, although these targeted units would not count to the minimum number of dwellings to be targeted since you are not allowed to mix and match targeted (biased) and random (unbiased) sampling. The inspector can also perform a worst-case inspection, which would target only units where the paint is deteriorated and young children are known to be present.

Once the dust and paint samples are collected, they are sent to a laboratory for lead analysis.

13.3.2.4 Evaluation of Analytical Results

Table 13.4 (Table 5.7 in the HUD guidance document) provides the hazard levels for a LBP risk assessment. Note that the maximum allowable levels are affected only by the concentration of lead in the paint. The steps that can be taken to control identified lead hazards are discussed later in this chapter.

TABLE 13.4

Hazard Levels for Lead-Based Paint Risk Assessments

Media	Level	
Deteriorated paint (single-surface)	5000 mg/kg or 1 mg/cm²	
Deteriorated paint (composite)	5000 mg/kg or 1 mg/cm² divided by the number of subsamples	
Dust (wipe sampling only)	Risk Assessment	Risk Assessment screen (dwellings in good condition only)
Carpeted floors	100 µg/ft²	50 µg/ft²
Hard floors	100 µg/ft²	50 µg/ft²
Interior window sills	500 µg/ft²	250 µg/ft²
Window troughs	800 µg/ft²	400 µg/ft²
Bare soil (dwelling perimeter and yard)	2000 mg/kg	
Bare soil (small high-contact areas, such as sandboxes and gardens)	400 mg/kg	

Note: mg/cm², milligrams per square centimeter; µg/ft², micrograms per square foot; mg/kg, milligrams per kilogram.

13.3.3 Designing the Lead-Based Paint Survey

Inspections for the presence of lead-based paint are outlined in Chapter 7 of the HUD guidelines. This section summarizes that chapter. Readers who want a more thorough description of this process or intend to become a certified LBP inspector, as required for all HUD investigations, are urged to read HUD Chapter 7.

LBP surveys are guided by two concepts: room equivalents and testing combinations. A *room equivalent* is "an identifiable part of a residence." It is analogous to a functional space as defined under the Asbestos Hazard Emergency Response Act (AHERA; see Chapter 10). The objective in defining a room equivalent is in identifying building areas with individual painting histories. The most common example of a room equivalent is a room (surprised?). Other types of room equivalents include hallways and stairways, common areas in multifamily buildings such as foyers and entranceways, and exterior areas such as playgrounds and walkways with painted railings. Closets are not considered to be room equivalents because they rarely have a painting history different than the room that they are associated with. Exterior sides should be considered the same room equivalent unless there are obvious dissimilarities.

Testing combinations consist of three components: a room equivalent, a building component type, and a substrate. Examples of testing combinations are provided in Table 13.5. Testing combinations are analogous to homogeneous areas in an asbestos survey by providing the basic unit of the LBP survey. It is important to note that color, the most conspicuous quality of paint,

TABLE 13.5

Examples of Testing Combinations

Room Equivalent	Building Component	Substrate
Master bedroom	Door	Wood
Master bedroom	Door	Metal
Kitchen	Wall	Plaster
Garage	Floor	Concrete
Exterior	Siding	Wood
Exterior	Swing set	Metal

is not part of the testing combination. The layer or layers of paint beneath the visible layer is more likely to be LBP. In fact, the deeper you dig into the paint layers, the more likely it is that LBP will be encountered. Therefore, any valid methodology to be used in an LBP investigation must evaluate the lowest layer of paint, that is, the layer of paint in contact with the substrate. As a rule of thumb, testing should be oriented to where the paint is thickest, since that area is most likely to include the most layers of paint.

13.3.4 Lead-Based Paint Inspection Methods

The goal of the lead-based paint survey is to determine the *lead loading*, also called area concentration, on a given surface area on a substrate. The simplest test method for the presence of lead in paint is the *chemical test kit*. These kits, available at most home improvement centers, provide an inexpensive way for the homeowner to test for the presence of LBP in the home. Typical test kits consist of swabs roughly the size of a small cigarette. These swabs contain a reagent that changes color when it comes into contact with lead. To test a painted surface, an incision is cut through the paint to the substrate. The reagent is then released from the swab by breaking the seal that holds the reagent in the swab, and the swab is rubbed over the incision, allowing the reagent to penetrate to the substrate. If any of the layers of paint contain lead, the reagent will change color. The drawbacks of this method are (1) the test is qualitative, and there is no way of knowing the lead content in the paint or whether the paint should be classified as lead-containing; (2) a failure to cut all the way to the substrate will result in failure to evaluate the layer or layers that are most likely to be LBP; and (3) the test is prone to false positives when other metals are present.

Paint chip analysis, described earlier in this chapter, is another method used to assess the presence of LBP in a residence. However, this test method is relatively time-consuming compared with other methods, the sampling method damages the painted surface, and part of the substrate sometimes detaches with the paint sample, which affects the lead concentration calculations.

The most reliable and common method of testing for the presence of LBP, and the method recommended by HUD and the USEPA, involves the use of an *x-ray fluorescence*, or *XRF machine*. The XRF machine contains a radioactive source (typically Cobalt-57 or Cadmium-109) that emits x-rays produced by radioactive decay.

When x-rays encounter lead, the lead gives off gamma rays at a certain energy level due to the photoelectric effect. These gamma rays enter an aperture in the machine, where they encounter a detector that records their presence. The XRF machine then converts the amount of gamma rays that have entered the machine into the concentration of lead in the tested surface. XRF testing should not be performed on paint that is chipped or peeling, since the XRF machine requires an even surface to provide an accurate reading.

A typical XRF machine is shaped like a gun (see Figure 13.5). The machine operator places the aperture at the end of the machine against the surface to be tested. When the machine operator presses the trigger, the chamber that contains the radioactive source is opened, and x-rays impinge on the paint surface being tested. The time that the aperture is opened depends upon the machine. Some instruments automatically adjust for source decay, that is, when the radioactive source begins to wear out.

XRF machines, because they contain a radioactive source, must be registered with the Nuclear Regulatory Commission. Their transportation across state lines is regulated, as is their usage and their disposal.

Because each machine will provide inaccurate readings if it is not used or maintained properly, it is important for the machine operator to follow the manufacturer's instructions on the usage of the machine. Integral to the operation of all XRF machines (and all quantitative analytical equipment, for that matter) is *calibration*.

Machine calibration is designed to ensure that the lead readings obtained are precise, in other words, perfectly reflecting the actual lead concentration in the paint tested. Calibration must be performed the first time a machine is used that day, after the machine has been transported (since vibrations and jostling during transit can affect the machine's operations), and at selected landmarks in the course of the testing day (for instance, every 4 hours or after a certain amount of tests are run).

Ideally, a calibration check will ensure that all lead readings are precise. In reality, the most important reading to ensure for accuracy is at the bright line between LBP and non-LBP, that is, 1.0 mg/cm^2. If a surface contains lead at a concentration of 12.0 mg/cm^2 and the XRF machine records the lead concentration at 8.0 mg/cm^2, it is less of a concern than if a surface contains lead at a concentration of 1.2 mg/cm^2 and the XRF machine records the lead concentration at 0.8 mg/cm^2. Figure 13.6 shows the results of a typical five-point calibration curve and correction. Machine calibration is performed before and after each inspection, and every 4 hours in the course of the inspection.

FIGURE 13.5
Using an XRF machine to test for lead-based paint. (Courtesy of Bay Area Lead Detectors Blog, www.bayarealeaddetectors.com/blog/.)

13.3.5 Testing Protocols

Once the XRF machine is calibrated, the survey begins. Testing protocol calls for at least one "shot" per testing combination. If, in the judgment of the investigator, a surface is sufficiently large that there might be variations in its painting history, then it should be tested more than once. The results of each test are recorded by the XRF machine, which then can be uploaded onto a computer. The investigator must be very systematic and organized in labeling the testing combination being tested. In a typical room, each wall must be designated in a way that the investigator can tell which reading refers to which wall. Investigators typically design their own methods of organizing and labeling the testing combinations to facilitate data interpretation once back in the office.

For instance, on the third floor of our fictitious building, shown in Figure 13.7, four cardinal directions have been assigned to each room in the building. Walls then can easily be referred to by their room designation and the appropriate direction. For instance, "303N" would refer to the wall in

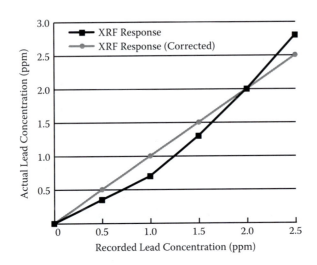

FIGURE 13.6
A typical five-point calibration curve.

Room 303 along the outside of the building. Whether north is accurately identified is less important than consistency and clarity in reporting the results.

13.3.6 Data Interpretation

Interpretation of the results of the XRF testing is guided by the *XRF performance characteristic sheet* provided by the manufacturer. One of the items specified on the XRF performance characteristic sheet is the "inconclusive range" of the machine. For instance, the sheet for one popular machine brand indicates that the inconclusive range is between 0.8 and 1.2 mg/cm². This means that all measurements falling into that range are statistically uncertain and cannot be used to determine the presence or absence of LBP on that testing combination. In such cases, the investigator can either (1) retest the testing combination; (2) confirm the results by collecting a chip sample and sending it to a laboratory for analysis; or (3) assume that the paint is LBP. As with all environmental concerns, one can always be conservative and assume the presence of environmental contamination. One is never allowed to assume the absence of environmental contamination, unless there is documentation to support that assumption.

The XRF performance characteristic sheet also provides information on how to calculate a *substrate correction*. Some substrates can interfere with the XRF measurements by either contributing x-rays or inhibiting them from entering the machine recording chamber. The XRF performance characteristic sheet will tell the user how to perform a substrate correction.

As an example, let's say that a reading from a testing combination is 1.1 mg/cm². However, the XRF performance characteristic sheet indicates that one

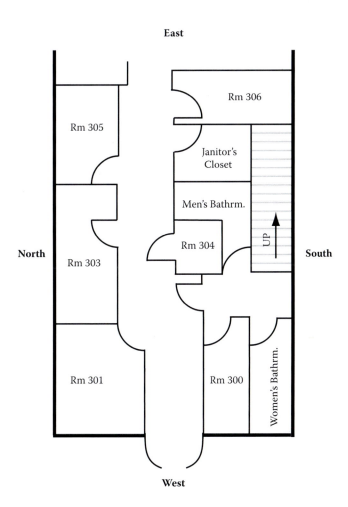

East

Rm 306

Rm 305

Janitor's Closet

Men's Bathrm.

North

Rm 304

UP

South

Rm 303

Rm 301

Rm 300

Women's Bathrm.

West

FIGURE 13.7
A diagram of an area about to undergo an LBP survey.

must subtract 0.5 mg/cm² for that substrate. Therefore, what first appeared to be a reading indicating the presence of LBP would now be 0.6 mg/cm², which indicates the absence of LBP.

To be certain that the appropriate substrate correction is being applied, HUD advises that in a single-family residence the investigator obtain three readings on bare substrate from two different locations. The substrate correction would be the average of the six readings.

13.3.7 Data Documentation

The final report should contain all of the relevant information about the method used in the investigation, the results of the investigation, and the investigator.

As with all environmental investigation reports, the lead investigation report should answer the where, when, why, who, and how questions. The qualifications of the investigator should be documented, and all test results should be reported, even test results that were inconclusive or otherwise unusable in the data interpretation. Labeling of the testing combinations should be sufficiently obvious that someone who had never been to the facility could figure out the locations of the LBP identified by the investigation.

Many jurisdictions have mandatory reporting requirements for LBP. All mandatory reporting should be made in accordance with the regulations, and the report should document that the required notifications were made.

13.4 HUD Guidelines for Lead-Based Paint Hazard Abatement

As with asbestos (see Chapter 12), a lead-based paint hazard can be abated either by enclosure, encapsulation, or removal. (Removal is the most invasive and usually the most expensive, and HUD suggests avoiding this abatement method.) One major difference between asbestos abatement and LBP abatement is that asbestos abatement is governed by regulations and LBP abatement is governed by guidelines, which nevertheless can be enforced by state and local authorities. This section describes the three active abatement methods.

13.4.1 Enclosure

Enclosure involves creating a "dust tight" barrier over the surface containing the LBP. HUD guidelines indicate that the barrier must be rigid and durable, able to last without replacement for up to 20 years. In the case of contaminated soils, the barrier could be a synthetic rubber surface commonly used on modern playgrounds. For exterior walls, it could be vinyl or aluminum siding; for interior walls, paneling or Sheetrock. As with all methods of abatement, a certified risk assessor must certify the integrity of the enclosure.

13.4.2 Encapsulation

Encapsulation involves coating the LBP-covered surface with either a liquid-applied coating or an adhesive material, such as wallpaper. The primary difference between enclosure and encapsulation is that for enclosure the barrier is rigid, typically applied with screws, nails, or other such materials, whereas with encapsulation, the material adheres to the surface without the aid of such materials. As with enclosures, the encapsulant should be designed to last for up to 20 years.

Encapsulation is not appropriate on highly deteriorated surfaces, where the encapsulant can delaminate along with the paint it is designed to protect, or high friction surfaces, where the longevity of the encapsulant is likely to be compromised. Paints can be used for encapsulation only if they are designed to withstand hazards associated with lead paint, such as resistance to chewing. The encapsulant manufacturer typically specifies application methods, appropriateness for various substrates, and testing procedures to verify the effectiveness of the encapsulation activities. In all cases, starting with a clean surface is key to the effectiveness of the encapsulation activities.

13.4.3 Lead-Based Paint Removal

LBP can be removed using different techniques than asbestos removal mainly because lead is not considered to be an airborne contaminant, or a hazard to older children and adults. Procedures for LBP removal are far less prescribed than asbestos removal. Various jurisdictions have regulations pertaining to LBP removal. In most areas, however, HUD guidelines are used in conjunction with OSHA regulations governing the protection of workers.

As with any environmental investigation or remediation, the first and most important rule is to do no harm. Disturbance of a hazardous material such as LBP releases lead into the air, where workers and residents can inhale it. Most precautions taken in an LBP removal project are designed to prevent releases from occurring.

LBP removal follows the steps of other types of remediation: design, implementation, and verification. The design portion of the LBP removal deals mainly with protection of residents and workers, establishing procedures for paint removal, and establishing protocols for verification of the LBP removal process.

13.4.4 Worker Protection

OSHA requires that workers wear proper personal protective equipment (PPE) to protect them from lead inhalation. As described at the beginning of the chapter, the ideal way to determine the level of respiratory protection for workers is to perform an initial exposure assessment by collecting personal air samples representative of a full shift for the job classification, in this case, lead abatement worker. In the real world, however, the employer, knowing that workers will be exposed to airborne lead in the course of the abatement activities, will supply its workers with respiratory protection. The respirator of choice for most LBP removal activities is half-face air-purifying respirators with high-efficiency particulate air (HEPA) filters.

Workers also wear protective suits, not to protect their skin, since lead cannot penetrate the dermal barrier, but so that dust can be removed from their persons to avoid cross-contaminating other portions of the building. The protective suits are removed as workers leave the work area and are placed in bags for eventual proper disposal.

13.4.5 Protecting the Residents

To protect residents, the area where paint removal activities occur must be unoccupied. Warning signs are posted, and openings leading outside the work area are isolated or protected. The level of work area protection is not nearly as elaborate and proscribed as for asbestos abatement, mainly because lead dust particles are not as aerodynamic as asbestos fibers and therefore are unable to travel as far or remain in the air as long. In addition, lead inhalation does not present the same threat to workers as does the inhalation of asbestos fibers, so the degree of protection needed for populations other than young children is less involved.

13.4.6 Preparing the Work Area

As with asbestos removal, there are several ways to prepare the work area. The level of protection to be implemented for a given LBP removal project will depend upon the following factors:

- The size of the surface(s) to be abated
- The type of hazard control methods to be used
- The extent of existing contamination
- Building layout
- Vacancy status of the dwelling

Prior to all work area preparation, dust and severely deteriorated paint should be removed (see Figure 13.8), since they are likely to be disturbed and possibly released outside the work area during the work area preparation activities.

HUD defines four levels of work area protection. *Level 1 protection* pertains to removal actions in which less than 2 square feet of LBP is to be disturbed per room. For a level 1 job, warning signs are placed outside the abatement area. One sheet of 6-mil polyethylene sheeting ("poly") is placed in the abatement area and extending 5 feet beyond the abatement area. Vents are sealed within 5 feet of the area, and furniture is covered within 5 feet of the area.

Level 2 protection is used when there is to be 2 to 10 square feet of LBP disturbance per room, or there is to be an interim control of a lead hazard while a permanent solution is being designed. These cases call for two layers of poly in the work area with an airlock on the doorway to the work area, as described in Chapter 12. All vents inside the room are to be sealed with poly and all furniture removed from the room prior to the onset of LBP removal activities.

Level 3 protection is used for a level 1 or level 2 project that will occur over the course of multiple days. For these projects, the residents are allowed to return to the dwelling at the end of each work day even though the lead abatement activities are still in progress. The rooms where lead abatement

FIGURE 13.8
Paint dust fallen from a wall in a residence.

is occurring must be firmly secured, and warning signs must be present at the main entranceway to the dwelling. The top contaminated layer of poly, which will be contaminated with lead dust and removed LBP, must be removed at the end of the work day. The guidelines also suggest the collection of dust samples outside of the area on an as-needed basis to verify that lead-containing dust has not escaped from the work areas.

Level 4 protection is used for large projects, that is, projects in which greater than 10 square feet of surface will be disturbed. For such jobs, there must be warning signs on the building exterior and airlocks on all doors.

Lead removal occurring on the exterior of a building can occur on exterior painted surfaces or by the removal of lead-contaminated soils. In these cases, a temporary fence or barrier tape is placed 20 feet around the surfaces or soils to be abated to restrict residents and others from the work area. Whether residents can return to their dwelling at the end of the workday will depend on the size and duration of the project. One layer of 6-mil poly is placed on the ground, extending to 10 feet beyond the surfaces or soils to be abated. Movable items within 20 feet of the surfaces or impacted soils are removed from the area, and entryways to the building within 20 feet of the surfaces or impacted soils are closed off. Warning signs are placed around the perimeter of the work area.

When removing paint from exterior surfaces, HUD recommends avoiding windy days to prevent the airborne spread of lead-containing dust or particulates in the course of the abatement activities.

13.4.7 Lead-Based Paint Removal Procedures

There are several approved methods for the removal of LBP. The simplest method is building component replacement, for example, taking off the door with LBP and replacing it with a new door painted with lead-free paint. Removing paint from a building component may involve scraping tools, heat guns, and paint strippers, although a paint stripper may pose an additional hazard to abatement workers if it contains compounds that are inhalation hazards to the workers. HUD prohibits any removal method that will create airborne lead-containing dust or gases, such as sanding, dry removal, or burning.

Gross removal of paint is followed by cleaning steps that involve HEPA vacuuming and wet washing. Vacuums must be supplied with HEPA filters since lead dust can be sufficiently small that they may not be trapped by conventional vacuum filters. For all lead removal methods, negative air pressure is not required because of the aerodynamics of lead dust. Simple water may be used in the wet washing process, although sometimes detergents are added to the water to improve cleaning.

For levels 1, 2, and 3 area preparations, one round of HEPA vacuuming and wet washing within 5 feet of the work area are sufficient. For level 4 area preparation, a "three-pass" cleaning process is used. In a three-pass cleaning, the workers will HEPA vacuum the surface, wet wipe, and then HEPA vacuum the surface again. The first HEPA vacuum removes most of the dust and dirt. The wet wash helps to dislodge the remaining dust and dirt from the surface, which then is picked up by the second round of HEPA vacuuming. A HEPA spray-cleaning vacuum can be used as an alternative to the three-pass cleaning system.

Once the LBP has been removed from the surface, the contaminated poly is placed in waste bags and the workers and their removal equipment are removed from the work area. At least one hour after completion of the removal activities, an independent third party inspects the work area. The one-hour time period allows for any dust to settle out of the air. If the inspector determines that the lead removal has been successful, then clearance sampling can begin. If there is evidence that lead remains in the work area, then the entire cycle of preparation, removal, cleaning, and inspection must be repeated.

13.4.7.1 Clearance Wipe Sampling

Unlike asbestos removal, clearance sampling involves the collection of wipe sampling rather than air sampling, since lead is less of an airborne contaminant than asbestos. It should be noted that XRF testing is not allowable for clearance sampling because of its difficulty in accessing the presence of dust on a surface. Wipe samples may be individual samples or composite samples. HUD recommends the following protocol for collection of individual clearance wipe samples:

- Collect wipe samples from window wells, window sills, and floors.
- Samples should be biased toward areas where a lead hazard was present or to high traffic areas.
- Two dust samples from at least four rooms in the dwelling should be collected and tested. These samples should be collected from the window wells, the window sills, and the floors. In common areas, there should be at least one wipe sample for every 2000 square feet of floor.
- For areas prepared with airlocks, a wipe sample should also be collected within 10 feet of the airlock to determine the effectiveness of the containment system.

If the inspector chooses to composite the clearance wipe samples, then three composite samples should be collected for every batch of four rooms. Only wipe samples collected from the same building components can be composited. Compositing must be performed in the field, not in the laboratory. The same rules (and limitations) apply for compositing clearance samples as investigation samples. The sampling frequency mirrors that for discrete clearance wipe sampling, as described earlier.

The following guidance values are used to determine whether the work area can be cleared for reoccupancy:

- Floors—40 μg lead/ft²
- Window sills—250 μg lead/ft²
- Window troughs and other rough surfaces—400 μg lead/ft²

If one or more clearance samples fail to meet the above standards, then the entire cleaning and testing process needs to be repeated.

Once finished, all porous surfaces should be coated with a sealant. Surfaces should be completely dry before the sealant is applied.

There are reporting requirements for various jurisdictions, and reoccupancy of the abated areas may require regulatory approval and possibly a sign-off by a licensed professional.

References

Dartmouth Toxic Metals Superfund Research Program, http://www.dartmouth. edu/~toxmetal/.

Getting the Lead Out: A Lead Prevention Timetable. n.d. *The Cooperator*, http://cooperator.com/articles/911/1/Getting-The-Lead-Out/Page1.html.

U.S. Department of Health and Human Services, Center for Disease Control. October 1991. Preventing Lead Poisoning in Young Children.

U.S. Department of Housing and Urban Development (HUD). 1997. Guidelines for the Evaluation and Control of Lead-Based Paint Hazards in Housing.

14

Drinking Water Testing

14.1 Introduction

Thanks to the Safe Drinking Water Act (SDWA), the water that runs from the tap in our homes, schools, and places of business is among the cleanest water in the world. The SDWA requires *public water systems*, defined as those systems that provide piped drinking water to at least 25 persons or 15 service connections for at least 60 days per year, to test the drinking water for a variety of contaminants on at least an annual basis (see Chapter 2). The water must be tested for various chemical, biological, and even nuclear contaminants, as well as physical properties, such as color, temperature, and hardness. Smaller drinking water systems may be regulated under state and local laws.

Then why does this chapter focus only on lead in drinking water? For one, lead is one of the most prevalent drinking water problems in this country. Most important, lead is often a building-specific problem, even for buildings serviced by public water systems but especially for those that have systems that are not monitored under the SDWA. Because the problem can be property specific, it is a real estate issue that can affect the sale and development of properties, which is why it is the only drinking water concern mentioned in the ASTM Phase I standard (see Chapter 5).

14.1.1 History of Lead in Drinking Water

Lead in drinking water dates back to ancient times. The chemical symbol for lead, Pb, is an abbreviation of the Latin word for lead, *plumbum*. Lead was used in the conveyance system that supplied drinking water to ancient Rome (see Figure 14.1). The English word "plumbing" derives from this Latin root. By its very origins, plumbing has been associated with lead!

Just as with asbestos (see Chapter 10) and lead (see Chapter 12) in paint, lead was utilized in drinking water systems because of its beneficial properties. Lead makes products more durable because of its resistance to physical as well as chemical breakdown, and it is relatively inexpensive to mine and

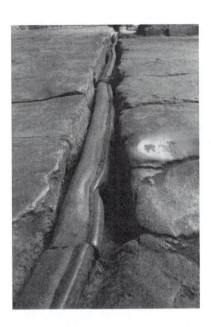

FIGURE 14.1
Lead piping at an ancient Roman bath.

manipulate. As demonstrated in Figure 14.1, lead pipes can serve its users for a long, long time.

Unfortunately, the leaching of lead into the drinking water, even in small quantities, can impact the health of the people drinking the water, particularly young children and pregnant women. Just as with the ingestion of lead-based paint, the ingestion of lead-contaminated drinking water can lead to neurological damage and even death if ingested in sufficiently large quantities.

Most lead gets into drinking water when the water comes into contact with plumbing materials that contain lead. This process is known as *corrosion*, which is defined as "the deterioration of a material, usually a metal, that results from a reaction with its environment" (NACE International, 2000). When the plumbing materials corrode, they will release lead into the water. Although corrosion occurs more readily when the water is acidic (low pH), or "soft" (low calcium), lead can leach into the water to a certain degree regardless of the physical properties of the water.

The SDWA established a maximum contaminant level (MCL) of 15 micrograms per liter (μg/l) for lead in drinking water (certain states have established lower acceptable limits for lead in drinking water). The SDWA also established an *MCL "goal"* (MCLG) of "non-detect" by conventional laboratory methods. The MCL is enforceable; the MCLG is not. Therefore, the MCL provides a bright line by which a determination can be made as to whether the drinking water in a particular building is fit for human consumption.

Starting in the 1800s and going well into the 20th century, lead was commonly used in water piping, solder, and plumbing fixtures (e.g., faucets, taps, elbows, and valves). Buildings constructed before 1930 were most likely constructed with lead-based piping, after which copper became the material of choice for water pipes. Between 1920 and 1950, galvanized pipes were also used for plumbing. The phasing out of lead in piping, however, did not extend to solder, which remained predominantly lead-based into the 1980s.

The 1986 amendments to the SDWA included the *Lead and Copper Rule*. This rule mandated a total ban of the usage of lead in piping (known as the *Lead Ban*), established a limit of less than 0.2% lead in piping solder and flux (the material used in association with the solder), and banned the usage of more than 8.0% lead in faucets and other plumbing fixtures. Prior to 1986, solder often contained as much as 50% lead, posing a significant hazard even though it was used in small quantities.

Despite the enactment of the Lead and Copper Rule in 1986, one cannot conclude that all piping, solder, and flux installed post-1986 complies with its regulations. The states had until 1988 to institute their own versions of these rules, and in many jurisdictions the rules were not enforced until 1991. Even that date is risky to rely on, since plumbers commonly used old materials stock or ignored the ban altogether for several years afterward. In fact, it was still permissible to sell piping and fittings containing lead in the United States until August 6, 1998. Therefore, it is advisable to consider the possibility of lead in the drinking water system for buildings or portions of buildings that were constructed well into the 1990s.

14.2 Sources of Lead in Drinking Water

To conduct a lead in drinking water investigation, one must first understand the potential sources of lead in a drinking water system. It is helpful to picture a drinking water system as having three major components: water withdrawal points, a delivery system, and access equipment.

Drinking water is withdrawn either from surface water or groundwater (see Chapter 4 to learn more about groundwater). Surface waters that provide drinking water are generally freshwater lakes, man-made reservoirs, and rivers, although many countries are now drawing much of their drinking water from saltwater bodies and processing the water through desalination plants.

Figure 14.2 is a schematic drawing of a water treatment plant that draws its water from a river. The water enters the plant through underground piping; is treated within the plant for various reasons, depending on the type of treatment required to make the water potable; and then piped to a storage facility, typically a water tower. The water tower releases the water on

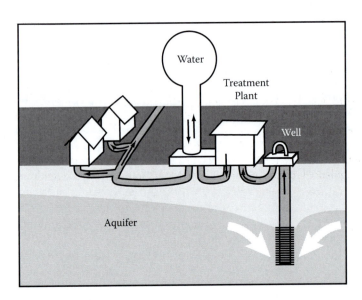

FIGURE 14.2
Surface water source of drinking water. (From Ohio EPA. With permission.)

an as-needed basis into underground piping, first through mains and then through a series of smaller diameter piping, until it reaches the buildings that will use the water. There is a service connection, typically at the roadside curb, that allows water to flow from the public water system into the building, where its thirsty inhabitants lie in wait.

Lead piping could be present in any and all of the various piping that comprises the delivery system. If the pipes are copper, then lead could be found in the solder that is used to connect the pipes and associated fittings.

Figure 14.3 is a schematic drawing of a water system for which groundwater is the drinking water source. Because the water source is below ground in Figure 14.3, drinking water wells must be installed in the aquifer. The well is outfitted with a screen, which allows the water to flow into the wellbore. A pump draws the water up to the surface and into the water treatment plant, after which the water follows the path described earlier.

In both systems, the equipment in the withdrawal piping, the equipment in the water treatment plant, and the delivery piping could contain lead. In the groundwater withdrawal system, lead also could be present in the pumps and associated equipment as well as the delivery piping.

Homeowners and businesses with their own private potable water systems typically get their water from private wells that draw groundwater from aquifers. The U.S. Environmental Protection Agency (USEPA) estimates that 15% of all homeowners have private wells on their property from which they draw water for potable purposes and for irrigation.

FIGURE 14.3
Groundwater source of drinking water. (From Ohio EPA. With permission.)

14.3 Lead in Drinking Water Investigation

14.3.1 Developing a "Plumbing Profile"

Before collecting samples and running analyses for the presence of lead, it is prudent to first develop a *plumbing profile* for the subject building. When was the building constructed? Have there been any building additions or renovations since its construction? If there was a gut (total) renovation or an addition added after 1986 (or whatever the applicable date in your state or territory), it is far less likely to contain lead in the plumbing system than if it was constructed before that date. Have new fixtures been added to the building during bathroom renovations and the like? How many drinking water outlets are present? The answers to these questions will affect the lead sampling plan.

It is always advisable to obtain lead data from the purveyor of the public water system if one services the building being tested. If the public water system has had problems in the past with elevated concentrations of lead, then the end users are likely to have lead problems as well. In some jurisdictions, the use of lead in service connectors was required prior to the Lead Ban. If the building has its own water source or is otherwise not connected to a public water system, then some due diligence into the results of previous water testing, if conducted, is in order.

Figure 14.4 shows the piping system on a floor of a typical commercial building. There are pipes carrying hot water and pipes carrying cold water.

Both systems have vertical pipes, called "risers," that distribute the water from its origin (usually in the lowest floor of the building) to the higher floors. In multistory buildings, bathrooms and other water withdrawal points tend to be in the same relative place on the floor because they connect to the same risers. Horizontal pipes that bring the water to the water outlets connect to the hot and cold water risers.

There are three areas of water withdrawal in Figure 14.4: the men's and women's bathrooms, and the janitor's closet. There may be multiple withdrawal points in the bathrooms, the withdrawal points of interest being the sinks. Inspection of the hardware, study of the systems, and their installation and maintenance history will determine how many withdrawal points to sample within the building. A visual inspection of the plumbing equipment can be

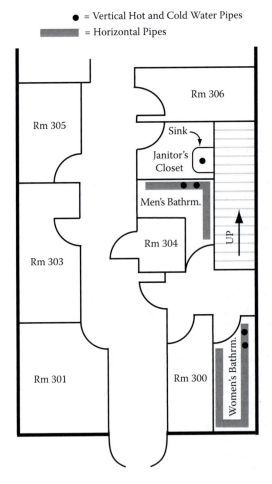

FIGURE 14.4
A schematic diagram of hot and cold water piping in a fictitious building.

useful. Lead pipes typically are dull gray in color and easily scratched by a steel object. If the equipment meets this description, it may be unnecessary to collect a water sample for analysis. Lead analyses are relatively inexpensive, so it pays to be thorough when performing a lead in drinking water survey.

14.3.2 Sampling for Lead in Drinking Water

Once the water outlets that may contain lead exceedances have been identified, it is time to collect water samples and have them analyzed for lead content. The first sample to be collected from an outlet is what is known as the *first draw* sample (see Figure 14.5). The USEPA recommends that the outlet remain unused for at least 6 to 8 hours prior to the collection of a first draw sample. This time lag would allow corrosion to take place, making the first draw water sample the worst-case scenario for lead in the drinking water at that outlet. Lead located anywhere within the water system will turn up in the first draw sample.

The containers to be used to store the water sample should be inert, that is, nonreactive with any chemicals in the water. Typically the containers are made of glass or laboratory-grade plastic. Analysis of the water samples is typically performed in a laboratory (Figure 14.6), although there are test kits available in various retail outlets. These test kits are not as accurate as a laboratory analysis but, like most test kits, are quick and inexpensive. Portable laboratories occupy a middle ground (Figure 14.7) due to their ability to analyze water samples quickly, albeit at a price similar to that at a fixed-base laboratory.

FIGURE 14.5
Collecting samples of tap water. (Copyright Agriculture and Agri-Food Canada, published by the Government of Canada.)

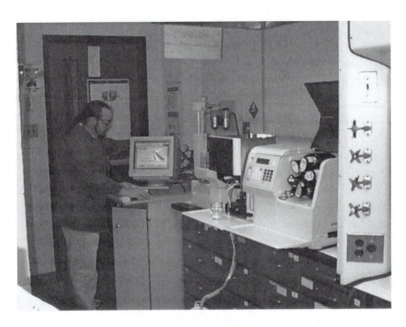

FIGURE 14.6
Laboratory analysis of drinking water sample. (From www.water-research.net. With permission.)

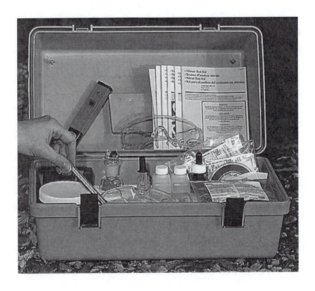

FIGURE 14.7
Portable laboratory for water analysis. (Courtesy of Lab Safety Supply, www.labsafety.com.)

If elevated concentrations of lead are detected in this sample, the next task is to determine whether the problem is with the water purveyor, the building, or both. There are several ways of doing this. One way is to collect a sample at the water's *point of entry* into the building. Easy access to the water's point of entry is relatively common for commercial buildings but less common for residential homes. If lead is detected above acceptable limits in the point of entry sample, then the problem lies with the water purveyor.

To determine if there is a lead problem at the fixture tested, it is necessary to discount the lead contribution from outside sources. For example, if 40 µg/l of lead was detected in the water emanating from a kitchen sink, and 15 µg/l was detected at the point of entry, then one can conclude that the sink equipment contributed 25 µg/l of lead.

A more complicated example would be if the water emanating from the kitchen sink contained 17 µg/l of lead and the water at the point of entry contained 10 µg/l of lead. In this example, one can conclude that the fixture contributed 7 µg/l of lead. Although the drinking water exceeds the limit, neither the source piping nor the fixture is by itself contributing lead over the regulatory limit.

Commonly, first draw water sampling is supplemented by what is known as a *flush sample*, which is collected from the same outlet from which the first draw sample was obtained. The objective of the flush sample is to purge the standing water from the outlet piping and allow fresh water from outside the fixture to pour through the spigot and be sampled. In small buildings, an adequate flush sample can be collected after letting the water run for 15 to 30 seconds. In larger buildings, the protocol is to let the water run until it gets cold, then sample it after it has run for 1 to 3 minutes, or even more in large buildings with complex water delivery systems. The investigator must use judgment to determine if the pipes have been sufficiently purged to have allowed water to pass through the point of entry, through the internal piping system, and to the water outlet.

Sampling results can be affected by numerous factors, so it is advisable to test multiple fixtures (in smaller buildings test all fixtures) before drawing conclusions about the sources of lead detected in the drinking water at one location.

14.4 Lead Mitigation

Once the lead problem has been identified, the next step is to mitigate (remediate) the problem. If the problem is exclusively with the water entering the building, then the solution may mean notifying the water purveyor of the problem and letting them fix it, which they will do by performing one or more of the actions described next. If the problem involves the water delivery system inside the building, then the building owner will have to take actions to protect the health of the building occupants.

Lead mitigation falls into two basic categories: passive mitigation and active mitigation. *Passive mitigation* is not the equivalent of doing nothing; it refers to a solution that does not involve engineering design and construction. For instance, allowing the water to run so that water that had been standing in the interior piping has been flushed may be sufficient to solve the problem. Samples could be collected after different flush times to determine how long the water would need to run before it can be consumed. This solution, while simple, wastes a lot of water and may not be advisable if young children are around, who cannot be held responsible for remembering to run the water before drinking it.

Another passive mitigation method is to not consume the water altogether and to rely on bottled water. Bottled water, however, can be expensive, especially if used for cooking as well as direct consumption. Because of the expense and inconvenience involved, bottled water generally is considered a temporary solution until a permanent solution can be implemented, except in situations where low consumption of water is expected, such as an office building.

The most reliable solutions to the lead problem involve *active mitigation*. The simplest solution, but usually the most costly, is to replace the offending delivery equipment with lead-free equipment. For a sink or a water fountain, this might mean simply replacing the faucet or perhaps the piping leading to the outlet. If, however, it involves removing pipes, which are typically concealed behind walls or beneath floors, then the mitigation could be very expensive and highly disruptive to building occupants.

If the water source is a nonpublic water system, then the building owner may be responsible for the entire system, not just the equipment inside the building. In such a case, the owner cannot blame the public water purveyor for the problem if the delivery system is to blame for the lead in the drinking water. In such cases, it may be advisable to install a treatment system between the water source and the outlets. Such devices are known as point-of-entry treatment (POET) devices.

One such POET system is a water distillation unit. A typical water distillation unit involves the process of *reverse osmosis*, shown on Figure 14.8. In such a system, lead is forced to move across a semipermeable membrane, going from an area of lower concentration to an area of higher concentration. Once on the other side of the semipermeable membrane, the lead can then be removed by flocculation or perhaps discharge to the sanitary sewer system. A reverse osmosis system can also be installed on an individual fixture, which would avoid the costs of replacing the plumbing hardware for that water outlet. Such systems at the tap are known as *point-of-usage* (POU) devices.

Another point-of-entry solution is to treat the water coming into the building so that it is less corrosive. This can be accomplished by adjusting the pH of the water coming into the building by treating it with devices such as calcite filters, soda ash, or phosphate solution tanks. Last, if the interior piping

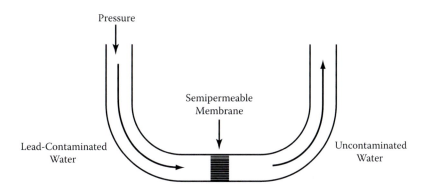

FIGURE 14.8
A schematic diagram showing the process of reverse osmosis.

is the problem, the pipes can be coated with a sealant that will prevent the water from interacting with the lead in the piping or solder, or a plumber can run a sleeve through the piping, which will also serve to prevent interactions between the water and the lead in the system.

The effectiveness of all engineered solutions must be verified by confirmation testing of the water at the locations where the lead problem was detected. Periodic testing, and maintenance of engineered systems such as POET devices, is also advisable to ensure that the lead problem, once solved, does not come back.

References

NACE International. 2000. NACE International Glossary of Corrosion-Related Terms. (Item No. 26012).

U.S. Environmental Protection Agency. June 1993. Lead In Your Drinking Water: Actions You Can Take to Reduce Lead in Drinking Water. EPA 810-F-93-001.

U.S. Environmental Protection Agency. April 1994. Lead in Drinking Water in Schools and Non-Residential Buildings. EPA 812-B-94-002.

U.S. Environmental Protection Agency. 1998. Commonly Asked Questions: Section 1417 of the Safe Drinking Water Act and the NSF Standard.

U.S. Environmental Protection Agency. January 2002. Drinking Water from Household Wells. EPA-816-K-02-003.

15

Mold Surveying and Remediation

15.1 Introduction

Mold is the most common of biological hazards for real estate. Part of the kingdom known as fungi, over 100,000 varieties of mold have been identified, and scientists estimate that another 100,000 varieties are yet to be identified. Molds can survive in the most hazardous of environments in the form of spores, which are shells that protect the mold from harm. The molds stay in their capsules until they find an environment in which they can thrive, at which point they shed their shells and begin their life.

Molds live comfortably in temperatures ranging from 40°F to 100°F (4°C to 37°C), the same temperature range as is found inside buildings. As with most living things, molds need oxygen, a nutrient source, and water to grow and amplify. Foods for molds, in general, are items with cellulosic content, such as wood products, paper products, and organic floor and wall coverings, and other common building materials. Once the mold spore encounters a moist area with available cellulose, it blossoms into a living mold, typically growing in circular patterns known as *colonies* (see Figure 15.1).

15.1.1 Health Impacts from Mold

Mold has been around humans for as long as there have been humans. Because of this long-term relationship, our species has adapted to being around mold and is immune from most of its health hazards. In fact, many molds, such as the mold from which penicillin is extracted, are beneficial to humans.

It is the molds that produce *mycotoxins* that create the health issues. There are over 100 molds (a minute fraction of all molds) that are known to produce mycotoxins, which are compounds produced by molds that are toxic to humans and animals. Mycotoxins enter the human body via the inhalation pathway, and then generally attach to the fat-rich organs in the human body, such as the liver or the kidneys. Temperature, pH, as well as the presence of other microbial agents, such as fungi, bacteria, and viruses, can affect the ability of a mold to produce mycotoxins.

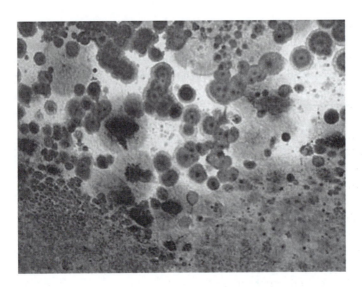

FIGURE 15.1
Mold growth as seen through a microscope. (Courtesy of Robert Kolodin.)

The health effects of mycotoxins do not follow a dose-response relationship, which is described in Chapter 7. Instead, health effects from exposure to mycotoxins vary significantly by individual susceptibility. Populations most impacted by mold include the elderly, people with allergies or asthma, or people who are immunocompromised. Adverse health effects can be acute, due to high, long-term exposure, or chronic, due to long-term exposure. The lack of a dose-response health-impact relationship has prevented the scientific and regulatory communities from developing permissible exposure limits and maximum permissible concentrations for toxic molds as has been done with chemicals and petroleum products.

Some of the more significant molds that produce mycotoxins are described next.

15.1.2 Types of Toxic Molds

The best known toxic mold is *Stachybotrys chartarum* (*atra*), or black mold. *Stachybotrys* favors building materials with high cellulose and low nitrogen content. Exposure to *Stachybotrys* can cause allergic rhinitis, skin irritation, sinus infection, pink eye, fatigue, and, in rare instances, brain damage and memory loss. There is evidence linking it to acute pulmonary hemosiderosis (bleeding in the lungs) in infants. *Stachybotrys* is not a hardy mold and usually fares poorly in the Darwinian competition for food and water. Consequently, the levels of *Stachybotrys* commonly found inside a building tend to be low compared with other molds, although that does not diminish its potential for harm. There are no known beneficial uses of any of the 16 species of *Stachybotrys*.

Aspergillus sp. is typically found in soil and originates from decomposing plant matter, household dust, building materials, and ornamental plants. There are 160 species of *Aspergillus*, of which 16 are toxic. Exposure to toxic forms of *Aspergillus* can cause sinus and local infections in vulnerable populations. These infections can be fatal in immunocompromised people. Certain species of *Aspergillus* produce a mycotoxin known as alfatoxin, which can cause cancer.

Penicillium sp. is commonly found in soil, food, cellulose, and grain. Some of the estimated 2000 species of *Penicillium* are well known for their medicinal value. Exposure to toxic forms of *Penicillium* can lead to skin, lung, and gut infections in immunocompromised people.

Other, less common forms of toxic mold include some species of *Fusarium*, *Trichoderma*, and *Memnoniella*.

15.1.3 Conditions Conducive to Mold Growth

Since molds are ubiquitous in all but the most arid portions of the United States, the key to controlling mold growth is not with the mold itself. Nor is it with the oxygen, which thankfully also is ubiquitous, or with the cellulose source, which is plentiful inside most habitable buildings. It is the presence of continuous moisture that leads to mold growth. Moisture is the most important issue to address when remediating mold.

Moisture usually derives from *water intrusion*. Water intrusion can be caused by flooding; damp conditions inside the building basement; leaks in one of the building's plumbing systems; or leaks in the roof or exterior walls of the building, as suggested by the water-damaged ceiling tile in Figure 15.2. It can also derive simply from the presence of high humidity, which can result from high temperature conditions inside the building, the presence of humidifiers, or the presence of improperly maintained commercial cooling devices. Relative humidity above 60% (the norm is 30% to 50%) will be conducive to mold growth.

15.2 Conducting a Mold Survey

There are several reasons for conducting a *mold survey*, which involves the identification of mold and conditions conducive to mold growth. Among the reasons to conduct a mold survey is in response to a health-related issue, reports of unpleasant musty or moldy odors, or the direct observation of mold. Another significant trigger for a mold survey is in support of a property transaction.

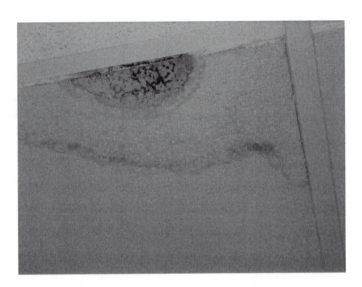

FIGURE 15.2
Mold growth on a water-damaged ceiling tile. (Courtesy of GZA GeoEnvironmental, Inc.)

There are no regulations regarding the investigation and remediation of mold, only guidelines. Among the guidelines commonly used are Canada's "Guide to Recognition and Management of Mold" (June 1995) and New York City's "Guidelines on Assessment and Remediation of Fungi in Indoor Environments," which was published in 2002. These documents are generally used by practitioners, although various aspects of these guidance documents are not universally accepted by the scientific community, as described in Section 15.4.3. Much of the ensuing description of a mold survey is taken from ASTM Standard E-2418-06, which ASTM developed in 2006.

The goal of the ASTM mold survey is to identify mold and conditions that are conducive to mold growth, such as deferred maintenance of a commercial building's building systems and moisture conditions. Since there are no nationally recognized certifications in the field of mold (except for Certified Industrial Hygienists), there are no certification requirements for the person or persons conducting any portions of the mold survey.

The three major steps in a mold survey are (1) a walk-through survey, (2) document reviews, and (3) interviews.

15.2.1 Walk-Through Survey

The walk-through survey is designed to inspect for evidence of mold growth or conditions conducive to mold growth. The investigator should inspect areas where mold is likely to be present, that is, areas where mold was reportedly seen or smelled, or areas where water intrusion is suspected or possible. Such areas should include places where water is known to be present, including basements, bathrooms, and kitchens; portions of the building near water bodies or septic systems; and areas with evidence of water damage or musty odors. In these areas, the inspector will look for evidence of mold growth on cellulose-containing surfaces, which include cardboard, paper, wallboard and ceiling tiles (both wallboard and ceiling tiles have paper surfaces front and back, with minerals and mineral fibers within). A thorough mold survey should also include an inspection of the heating, ventilation, and air-conditioning (HVAC) system.

Some building materials, such as dry wall covered with vinyl wallpaper, carpet, or wood paneling, may act as vapor barriers, trapping moisture underneath their surfaces and thereby providing a moist environment where mold can grow. Therefore, the mold survey should also include an inspection of "hidden areas," including crawl spaces, utility tunnels, and the tops of ceiling tiles. Inspection of these hidden areas must be balanced with consideration to the building owner and tenants, since some of these hidden areas can only be accessed by damaging the building materials. A direct search for moisture in unobservable areas can be performed using a moisture meter. *Moisture meters*, which come in many varieties, will record the amount of moisture in a building component. *Borescopes* are used to physically look into spaces not otherwise visible to the inspector, such as the interstitial space between wallboard units and crawl spaces. Access to the interstitial space between wallboard units can be obtained by the removal of lighting switch plates or electrical outlet plates. Borescopes are sometimes used in conjunction with wall check sampling devices, which enable the inspector to collect bulk samples of suspect mold growth. Remote moisture meters are also used to measure moisture behind walls (see Figure 15.3).

The scope of the mold survey needs to be understood by all parties prior to the onset of field activities. Surveys triggered by a health-related issue usually will be more comprehensive than a survey triggered by, say, a property transaction. According to the ASTM mold standard, the person conducting a mold survey in support of a property transaction, known as the *field observer*, does not need to conduct an "exhaustive" survey. Only "representative" offices in an office building or apartments in a multitenant building need to be inspected in the absence of a report of a mold-related issue. From these observations, the field observer, or the person in charge of the survey if a project team is involved, must extrapolate to make a determination regarding the presence of mold in the uninspected portions of the property.

Similar limitations that apply to an ASTM-compliant Phase I Environmental Site Assessment apply to the mold survey, such as "readily accessible," "reasonably available," and the like.

15.2.2 Document Review and Interviews

As with all environmental investigations, review of relevant documents and interviews with knowledgeable parties enables the investigator to get

FIGURE 15.3
Measuring moisture behind a wall using a moisture meter. (Courtesy of GZA GeoEnvironmental, Inc.)

a handle on the facts pertaining to the issue at hand. For mold, document reviews and interviews will provide information regarding mold-related health issues among building occupants, past instances of water intrusion, and building maintenance issues that might have resulted in water intrusion that would favor the growth of mold. When health issues are involved, these two steps are critical, and should be performed before the walk-through survey is conducted, so the walk-through survey can be biased toward the areas of concern.

Documents, available from the building owner, building tenant, or the local health department, will provide important information pertaining to the mold survey, especially when a health-related incident has occurred. Useful documents may include prior mold and water intrusion surveys; indoor air quality reports; health department inspections, violations, and so forth, which may be related to mold conditions; and engineering reports, especially structural engineering reports or property condition assessments, which may discuss water intrusion issues.

Whenever possible and as applicable, the people with mold-related symptoms should be interviewed. Such interviews can be tricky, since allergic symptoms can appear for a variety of causes in addition to the presence of mold. Assessing the timing of events and the location of things are critical. Did the symptoms appear before or after the basement became flooded? People with knowledge of the building and its systems should be interviewed as well.

15.3 Mold Sampling and Analysis

Unlike other types of environmental investigations, mold sampling is not required and often not even needed. Seeing the mold is conclusive evidence of its existence, and one can move right to the remediation step without the time and expense of collecting samples. Often, however, there may be reasons to collect mold samples even when the presence of mold is obvious, such as in support of a litigation or as part of the medical evaluation process.

Sampling methods for mold fall into two general categories: bulk sampling and air sampling. Both sampling methods are discussed next.

15.3.1 Bulk Sampling Methods

There are four types of *bulk sampling* methods.

- *Gross sample collection* involves using a hand tool to scrape or cut a piece of the suspect material, which is placed in an airtight bag and sent to a laboratory for analysis.
- *Contact or tape sampling* involves pressing a sticky material, such as tape, onto the surface suspected of containing mold. This sampling method, like gross sample collection, requires visual confirmation of the presence of mold to be effective.
- *Wipe sampling* involves applying a gauze pad or equivalent onto a measured portion of the surface suspected of containing mold (see Figure 15.4).
- *Microvac dust vacuum sampling* uses a microvacuum to collect dust samples for laboratory analysis. This equipment used in this process is similar to air sampling (see next section). It is the only bulk sampling method that does not require visual confirmation of the presence of mold. This sampling method is typically employed when it is unfeasible to make visual confirmation due to existing conditions, such as the presence of obstructions preventing visual inspection for mold.

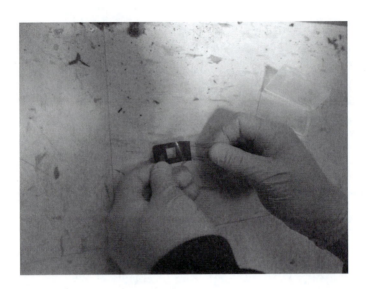

FIGURE 15.4
Using tape to collect a bulk mold sample.

Two types of analysis can be performed on bulk samples: viables analysis and nonviables analysis. *Viables* are molds that are living or have the potential to live, and are incubated in a laboratory prior to analysis. *Nonviables* are molds that have died, currently exist as spores, or do not have the potential to live. They do not require incubation, and therefore can be analyzed quicker and more cheaply than viables. The tradeoff with nonviable analysis is that nonviables give a limited view of the actual risk posed by the molds, since the health risks involved cannot be determined without knowing whether the molds are alive or can live. The results of a nonviables analysis include all molds, both viable and nonviable.

Table 15.1 provides a summary of nonviable analyses of bulk samples collected inside a house. The results, shown in *mold counts per gram*, indicate that mold is present at the four sampling locations (which is not a surprise) but is far more prevalent on the north wall of the family room. Since there are no standards for the presence of mold, all four detections can be considered significant; their significance is left to professional judgment. The molds detected are not associated with mycotoxins.

Table 15.2 shows the analytical results for viable mold analyses, measured in *colony forming units (CFUs) per meter.* CFUs are used since each mold spore can grow into a colony, forming the characteristic circular mold growth pattern shown in Figure 15.1. The quantities are generally less than the larger quantities shown in Table 15.1 and do not suggest a huge infestation of mold. On the other hand, some of the mold genuses shown in the table are associated with mycotoxins. *Speciation* (identifying which species within the genus

TABLE 15.1

Total Mold Analytical Results

Location	Results (Count/Gram)	Primary Fungal Spores
Header over door, family room	78,200	*Cladosporium*, amerospores
North wall of family room	2,300,000	*Cladosporium*
Bulk insulation at western roof	13,835	*Cladosporium*
Bulk insulation in attic	9384	*Cladosporium*, *Nigrospora*

TABLE 15.2

Viable Mold Analytical Results

Location	CFU/m	Primary Viable Molds
New kitchen area	966	*Cladosporium*, *Penicillium*, *Fusarium*
Basement crawl space	530	*Penicillium*
Basement office area	2097	*Penicillium*, *Cladosporium*
Master bedroom	141	*Aspergillus*, *Penicillium*
Outside background	3588	*Cladosporium*

are present in the sample), could indicate whether any of the species detected are associated with mycotoxins.

15.3.2 Air Sampling for Mold

Because mold spores are everywhere, air sampling for mold spores cannot conclusively distinguish between normal background presence of spores and the presence of mold spores due to excessive mold growth. Also, as with any type of air sampling, mold air samples provide only a snapshot of the moment in time in which the sampling occurred. They do not take into account mold variations by season, by time of day, or by temperature or relative humidity. Due to these limitations, air sampling is not part of a routine mold assessment. On the other hand, air sampling for mold can be useful in locating mold that could not be visually located or bulk sampled, or in support of a health-related investigation or a mold-related litigation. Pre- and postremediation sampling can help determine if mold remediation efforts have been effective, although air sampling is not part of remediation protocols, as discussed in Section 15.4.

Selecting the locations to be sampled is subjective and must be decided on a case-by-case basis. At a minimum, it is usually proper to collect at least two samples indoors—one in the area of interest and one near the area of interest—to provide information on the potential spread of mold through the building. Because mold is in the ambient air, an outdoor air sample also must be collected and used as a baseline for comparative purposes.

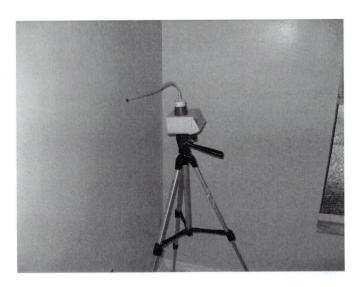

FIGURE 15.5
Collecting an air sample from behind a Sheetrock wall. (Courtesy of Robert Kolodin.)

Collecting an air sample for nonviable mold analysis involves using an air pump (see Figure 15.5), which is attached to an *impact cassette*, a cassette that is equipped with a sticky surface that will capture the mold, mold spore, or other *bioaerosol* (a biological agent that can be airborne, such as mold spores, viruses, and pollen) upon impact when drawn into the cassette. The collection method calls for running the air pump at 15 liters per minute (l/min) for up to 10 minutes. A wall check sampling device connected to an air pump can enable the collection of air samples from within wall cavities. The cassette is sent to a laboratory, where the sticky surface is removed, stained, and analyzed under a microscope. As with bulk samples, air samples can be analyzed either for viable mold or nonviable mold. The laboratory will normalize the analytical results to a standard based on the air pump flow rate and time used to collect the sample.

To collect an air sample for viable mold analysis involves *liquid impingement* or *gravitational sampling* techniques. Under the liquid impingement technique, an air pump pulls bioaerosols into a liquid collection medium, which is then placed onto nutrient agar. The gravitational technique involves using an air pump to send bioaerosols impinging through an impactor designed to hold an agar growth medium. The sample collection method calls for running the pump at 28.3 l/min for up to 10 minutes. The agar plate is then sent to a laboratory to incubate for a couple of days prior to analysis for viable mold.

For all air sampling methods, sample analysis should follow analytical methods recommended by the American Industrial Hygiene Association

TABLE 15.3

Air Sampling Results for Total Mold

	Room 303		Room 305		Outdoor	
Spore Type	Raw Count	Count/m³	Raw Count	Count/m³	Raw Count	Count/m³
Alternaria	0	0	0	0	2	88
Cladosporium	3	132	4	176	25	1100
Penicillium	2	88	2	88	8	352
Stachybotrys	2	88	0	0	0	0

(AIHA), the American Conference of Governmental Industrial Hygienists (ACGIH), or other professional guidelines.

15.3.3 Interpretation of Air Sampling Results

An example of mold air sampling results is provided in Table 15.3. In this table, the analytical results are normalized to *counts per cubic meter of air*. Of significance in Table 15.3 is the detection of *Stachybotrys* in Room 303, as well as *Penicillium*, some species of which can produce mycotoxins. It should be noted that only two actual *Stachybotrys* mold or mold spores were detected in Room 303. This is a small number, which does not differ statistically from a reading of 1 or 4, and therefore runs the risk of being misinterpreted. However, the presence of any *Stachybotrys* at all is of concern, especially since *Stachybotrys* was not detected in the outdoor baseline sample.

15.4 Mold Remediation

As stated earlier, there are no regulations regarding mold remediation, only guidelines. The most widely used guidelines are described in this section of the chapter.

15.4.1 Worker Protection

The lack of a dose-response health relationship has made it difficult and controversial to establish guidelines to protect workers from exposure to mold. There are, however, known health-related issues that can affect mold remediation workers even if they are not in one of the categories of susceptible humans. Acute exposure to mold, known as *organic dust toxic syndrome* (ODTS), can affect workers performing mold remediation. A more chronic mold-related illness is *hypersensitivity pneumonitis* (HP), which results from repeated exposures to the same toxic mold and therefore a concern to mold remediation workers. For these reasons, and for the purposes of conservatism, the U.S. Occupational Safety and Health Administration (OSHA) recommends that remediation workers use respiratory protection and protective suits when performing mold removal (see Figure 15.6), as discussed in Chapter 6.

15.4.2 Remediation Methods

In most cases, the best method for remediating mold-impacted building components is removal and replacement, and, most importantly, elimination of the source of moisture. The use of *biocides,* such as chlorine bleach, on the mold-impacted areas will kill the mold; however, dead mold can still be toxic, since the mycotoxins may still be present, lying in wait to do their damage. Therefore, it should only be used in limited circumstances, most important when a building's structural components, which cannot be safely

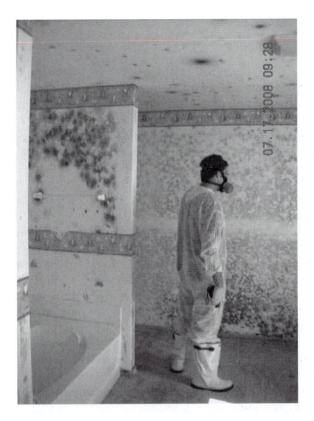

FIGURE 15.6
Mold removal worker wearing a protective suit and respiratory protection. (Courtesy of GZA GeoEnvironmental, Inc.)

or inexpensively removed, are affected. Simple scrubbing with water detergent will remove mold from hard surfaces, such as resilient flooring. Wet vacuums should be used to remove excess moisture from porous building components, carpeting, and furniture. The final cleanup should include drying the remediated area, then using a high-efficiency particulate air (HEPA) vacuum to remove remaining debris. As with asbestos (see Chapter 12), an encapsulant can be utilized postremoval to lock in the remediated surface or as a stand-alone remediation method, especially for situations where removal is impractical, such as for structural components.

HVAC systems impacted by mold present a special problem, since the mold will spread throughout a building as long as the HVAC system is contaminated. Remediation of a contaminated HVAC system involves shutting down the source of the water into the system and removal of cellulosic materials inside the HVAC system (which should not be there in the first place, and whose presence probably indicates that the system is malfunctioning or poorly maintained). Biocides can be used as recommended by manufacturer.

15.4.3 Preparing the Work Area, Differing Guidelines

There is some controversy in the scientific community whether projects should be defined by their size in lieu of meaningful dose-response data. Nevertheless, guidelines published by the New York City Department of Health and Mental Hygiene (DHMH) are widely recognized and used and are recommended by OSHA. Guidelines established by the U.S. Environmental Protection Agency (USEPA) are also widely used. Both sets of guidelines are described next.

15.4.3.1 New York City DHMH Guidelines

The DHMH guidelines establish four project levels, based on the quantity of mold-impacted building components to be remediated, and establishes a separate project level for the remediation of HVAC systems in commercial buildings. Many of the materials and procedures described here are intentionally similar to materials and procedures used in asbestos abatement, as described in Chapter 12.

A *level I remediation project* deals with less than 10 square feet (SF) of impacted building components. Level I remediation can be performed by maintenance personnel who are trained to use respiratory protection and wearing half-face respirators (minimum). The work is performed in an unoccupied area, without containing the work area. Standard dust suppression methods are used, and materials that cannot be adequately cleaned are placed in airtight, 6-mil polyethylene ("poly") bags and sent off site as ordinary waste. The project is completed when the remediated area is dry and visibly free of contamination and debris. No clearance testing is required for a mold remediation project.

A *level II remediation project* covers the remediation of 10 to 30 SF of impacted building components. A level II project differs from a level I project in that it requires the placement of plastic sheeting over surfaces that could become contaminated in the course of the mold remediation activities.

A *level III remediation project* covers the remediation of 30 to 100 SF of impacted building components. A level III remediation project differs from a level II project in that it requires the use of workers experienced in mold remediation, and requires the sealing of ventilation ducts and grills in the work area to prevent the cross-contamination of other portions of the building during the mold remediation activities.

A *level IV remediation project* is a large remediation project (greater than 100 SF in area). On level IV projects, workers should wear a full-face respirator, protective clothing, and gloves. The work area should be completely isolated prior to the onset of remediation activities and placed under negative air pressure with the usage of an exhaust fan equipped with a HEPA filter (see Figure 15.7). Airlocks and a decontamination unit should also be present to prevent the remediation activities from cross-contaminating nearby areas (Chapter 12 has a detailed description of a construction of a decontamination unit).

15.4.3.2 USEPA Guidelines

The USEPA's "Mold Remediation in Schools and Commercial Buildings" (March 2001) defines three levels of mold remediation projects: small (<10

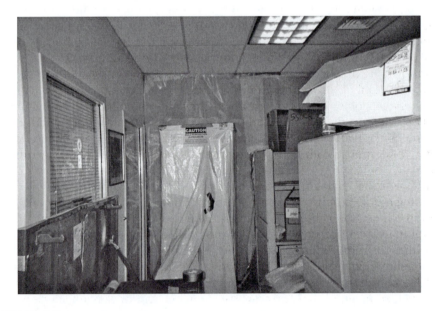

FIGURE 15.7
Entrance to a fully contained mold abatement work area. (Courtesy of GZA GeoEnvironmental, Inc.)

SF), medium (10 to 100 SF), and large (>100 SF). Remediation procedures and worker protections associated with these mold removal projects are similar to the New York City and OSHA guidelines. However, preparation of the work area for projects larger than 10 SF differs from the DHMH guidelines.

For medium-sized projects, the USEPA recommends the use of a "limited" containment, which consists of an enclosure consisting of one layer of 6-mil poly around the work area and sealing of air vents, doors, and other openings to outside the work area. Negative pressure, established using HEPA-filtered air units, is required within the containment area. "Full" containment is recommended for large remediation projects. Full containment entails the use of double layers of poly sheeting rather than a single layer as specified for a level IV remediation project. Otherwise, the work area preparation is consistent with a level IV remediation project as described in the DHMH guidance document.

References

ASTM International. 2006. Standard Guide for Readily Observable Mold and Conditions Conducive to Mold in Commercial Buildings: Baseline Survey Process. E-2418-06.

American College of Occupational and Environmental Medicine. 2002. Adverse Human Health Effects Associated with Molds in the Indoor Environment.

American Industrial Health Association. 2004. AIHA Guidelines on Assessment, Remediation and Post-Remediation Verification of Fungi in Buildings.

Canada Health Department. June 1995. Guide to Recognition and Management of Mold.

GZA GeoEnvironmental, Inc. 2003. Mold Identification, Recognition, Measurement, Toxicity and Abatement.

National Clearinghouse for Worker Safety and Health Training. 2005. Guidelines for the Protection and Training of Workers Engaged in Maintenance and Remediation Work Associated with Mold.

New York City Department of Health. 2002. Guidelines on Assessment and Remediation of Fungi in Indoor Environments.

U.S. Environmental Protection Agency, Office of Air and Radiation, Indoor Environments Division. March 2001. Mold Remediation in Schools and Commercial Buildings. EPA 402-K-01-001.

U.S. Occupational Safety and Health Administration. 2003. A Brief Guide to Mold in the Workplace.

16

Radon Surveying and Remediation

16.1 Introduction

As lead is to drinking water, radon is to indoor air, in that it is the most common of the pantheon of indoor air quality problems, and therefore merits its own chapter. Radon, like lead, is an element, but that is where the similarities end. Lead was known by the ancients and has been valued over the millennia for its strength and stability. Radon, discovered in 1900 (Partington, 1957), is a gas at ambient temperatures, does not react with other elements (is a "noble gas"), and has no beneficial uses. Radon is hazardous to humans because it is *radioactive*, emitting an alpha particle within what is known as the "uranium decay series" (see Figure 16.1). That alpha particle, when inhaled, can damage the delicate tissue deep inside the lungs, which increases the likelihood of lung cancer. The U.S. Environmental Protection Agency (USEPA) and the Surgeon General's Office rate radon as the second leading cause of lung cancer in this country after cigarette smoking (USEPA, 2009).

Although people might think that uranium is only found in areas where it is mined, traces of uranium are present in numerous rock formations throughout the United States. Buildings constructed on top of rock formations that contain trace concentrations of uranium have the potential to develop a radon problem. The USEPA estimates that 1 in 15 homes in the United States have elevated concentrations of radon.

What makes radon a problem specific to building interiors is its mobility as a gas. As shown on Figure 16.2, radon atoms can work their way into a building through cracks in the foundation or walls; gaps between construction joints or suspended floors; and openings that are part of the building structure, for instance, floor and basement wall penetrations designed to connect the building to outside sewer, electric, telephone, and natural gas utilities. Radon can even enter a building via the water supply, where it can be ingested directly into the body or inhaled through the shower or other airborne water exposure routes. Radon's entry into the building is facilitated by the building structure and the lower vapor pressure inside the building compared to the underlying soil or rock. Once inside the building, the radon concentration can build up, especially if there is poor ventilation, such as

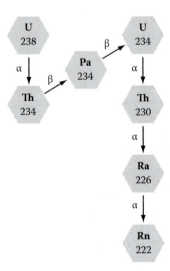

FIGURE 16.1
The Uranium-238 decay series leading to the formation of radon.

usually occurs in the basement of a residence, or there is otherwise little exchange with the outside air, as usually occurs when the house is sealed in times of extreme temperatures (summer or winter).

Radon cannot be smelled, tasted, or seen, making it an insidious threat to building inhabitants. The USEPA has set a limit for radon inside buildings at 4.0 *picocuries per liter of air* (pC/l), although it recently recommended lowering the radon action level from 4.0 pC/l to 2.0 pC/l. A picocurie is an extremely small number, equal to about the decay of two radioactive atoms per minute. This concentration forms a bright line by which radon hazards inside a building are assessed.

Millions of homes in this country have been tested for radon. Because of this large database, and because radon is typically related to the geology of the underlying rock and soil, states have been able to develop databases of radon potential by county and in many states by municipality. Following the USEPA protocol, areas are rated as zone or Tier 1 (high radon potential), zone or Tier 2 (moderate radon potential) or zone or Tier 3 (low radon potential). Figure 16.3 shows the radon potential for counties in the United States.

Under no circumstances should these designations be interpreted as meaning that a radon hazard is or is not present inside a building. They are merely indicators of the potential for a hazard to exist. The only way to determine whether a radon hazard is present is to run tests inside the building. With that stated, the tier ranking is often used in due diligence situations as a guide as to whether a radon survey is warranted. A Tier 1 ranking is more

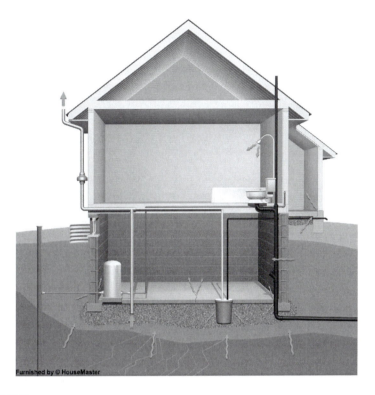

FIGURE 16.2
A diagram showing various paths through which radon enters a building. (Courtesy of HouseMaster Home Inspections. Housemaster® is a registered trademark of DBR Franchising, LLC.)

likely to generate a radon survey as a follow-up to a Phase I ESA than would a Tier 3 ranking, for instance.

16.2 Investigating for Radon

Radon can be tested using active devices or passive devices. *Active radon testing devices*, which can run on batteries or alternating current, continuously measure and record the amount of radon or its decay products in the air. These devices can provide a data report that can provide time-specific information on the radon condition inside a building. These devices tend to be far more expensive than passive screening devices, and require a qualified operator to run and maintain the equipment. Active radon testing devices are generally used in industrial settings where uranium or related radioactive elements are being handled.

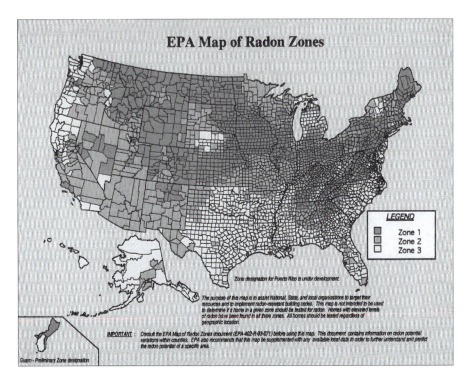

FIGURE 16.3
A map of radon potential zones in the United States. (From the U.S. Environmental Protection
Agency, www.epa.gov/radon/images/zonemapcolor_800.jpg.)

Far more common are so-called *passive radon testing devices*. They are called
passive because they do not require a power source to operate. Typical passive devices include charcoal canisters, alpha-track detectors, charcoal liquid scintillation devices, and electret ion chamber detectors (see Figure 16.4).
These devices are widely available online and at stores that cater to do-it-
yourself homeowners. They typically are small enough to fit into the palm of
a hand and are black in color so that they can be placed discreetly in various
portions of a building. Within the canisters are chambers that are designed
to either collect radioactive particles or record the change in ionization to the
collection media caused by the radioactive particles. Once activated for the
designated period of time, they are packaged and sent to a laboratory that
will calculate the radon level based on the canister contents.

The USEPA recommends testing for radon in the lowest level of a building
"suitable for occupancy." This is because the lowest level of a building has
the maximum contact with the underlying soil and rock formations, and is
closest to the penetrations or cracks through which the radon is entering the
building, hence the area most likely to contain elevated concentrations of
radon. The level needs to be suitable for occupancy because the radon limit

FIGURE 16.4
A passive radon measuring device. (From the National Institutes of Health.)

set by the USEPA assumes a certain residence time in that area to obtain the necessary dosage of radon.

Radon testing should be biased to simulate worst-case occupancy conditions at the level to be tested. The worst-case conditions should be simulated by sealing the room by closing the doors and windows, and disabling all air exchangers such as air conditioners at least 12 hours before conducting the test. That said, the test should be conducted under normal heating and cooling conditions; the environmental professional will have to balance these priorities in buildings located in hot climates. The canisters should be placed in areas that are not drafty and placed in or near the breathing zone to simulate conditions for the building occupant. They are typically kept in place for a minimum of 48 hours; the canister manufacturer will specify the recommended residence time for its product.

Weather conditions in the course of the test should be noted. Low barometric pressure, high wind conditions, and storms could affect the test results.

It is often tempting to place the canisters in areas where people are unlikely to reside, such as a boiler or laundry room, because of the decreased likelihood for interference, tampering, and traffic in and out of the room. This is only acceptable if there is reason to believe that the test in that room will yield similar results to a test that would be run in a room suitable for occupancy. The regulatory limit of 4.0 pC/l is based on risk data that assumes that the subject person is living at that level of the building. If a typical building resident spends no more than an average of 2 hours per week in the basement of a house to, say, do the laundry, the test will not do an adequate job of simulating actual resident exposure to radon gas.

In most scenarios, more than one radon canister will be employed within a building. This practice avoids the happenstance of a poor test due to a defective or damaged canister, or due to interference or tampering by a building occupant. Canisters should also be placed in more than one room at the desired building level so that spurious results from unusual building or geologic conditions could be identified by the person interpreting the data. It is also advisable to collect radon information from the next higher level, so that if there are exceedances of radon levels, they can be fully understood before implementing a remedy.

Large buildings with multiple levels and multiple rooms on each level pose a challenge to the environmental professional, since it may be impractical or costly to perform a radon survey in every room suitable for occupancy. In such situations, some forward investigation may guide the formulation of the survey. For instance, if certain rooms have more penetrations or more foundation cracks than other rooms, a survey of those rooms could be viewed as a worst-case scenario. In other words, if those rooms pass the radon test, then the investigator can justifiably assume that rooms with less propensity for the presence of radon gas would pass as well. If there are known geological variations beneath the building, however, testing should be done in rooms overlying all known geological strata, since that can affect the quantity of source material for the radon and therefore the amount of radon available to enter the room.

Although there is a bright line of 4.0 pC/l for assessing a radon hazard, the USEPA does recognize that natural variations and test variations can affect the results of a radon test. The USEPA's "Home Buyer's and Seller's Guide to Radon" indicates that a result of 4.1 pC/l has a 50% chance of actually being under 4.0 pC/l. Borderline results (including borderline results below 4.0 pC/l) should be rerun to ensure that the radon hazard is accurately assessed.

16.3 Radon Mitigation

The simplest way to address a radon hazard within a building is to install a radon mitigation system inside the building before it is constructed. This statement seems to beg the question as to how you would know the radon hazard exists, since radon tests cannot be run inside a building that has not yet been constructed. However, few buildings are built nowadays in untamed wilderness. Almost all buildings constructed in the United States are built near other buildings, and if not, built on geologic formations on which other buildings have been constructed. Therefore, it is often fairly straightforward to anticipate when the threat of a radon hazard might be present for a building to be constructed. Because it is so much less expensive to build a radon mitigation system for a new building rather than retrofit a mitigation system into an existing building, it is advisable to err on the side

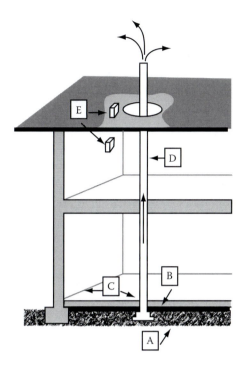

FIGURE 16.5
Diagram of a postconstruction radon mitigation system. (Adapted from the U.S. Environmental Protection Agency.)

of conservatism and design the new building with a radon mitigation system if there is reasonable doubt about the radon hazard on the property.

A typical radon mitigation system is shown on Figure 16.5. A sub-slab gas-permeable layer (A) is placed below the building foundation, which acts as a low-pressure bathtub for the collection of gases beneath the building slab. It is covered with plastic sheeting (B), and utility penetrations are sealed and caulked (C), thus preventing uncontrolled entry of gases into the building. A vertical vent pipe (D) is installed to channel gases from beneath the building slab to the roof, where a junction box (E), which operates a small fan, vents the gases safely into the atmosphere. As for drinking water, a breathable air *point-of-entry treatment system*, similar to the POET system described in Chapter 14, will control the entry of radon into the building via that pathway.

Figure 16.6 is a diagram of a postconstruction radon mitigation system. Because the subfoundation engineering controls cannot be retrofitted into an existing building, multiple penetrations beneath the slab are warranted. Pressure tests in the lowest level of the building will indicate the extent of the radius of influence a vacuum will have in pulling vapors from beneath the slab, which will dictate the number of penetrations needed for an effective radon mitigation system. This process is similar to the process of vapor

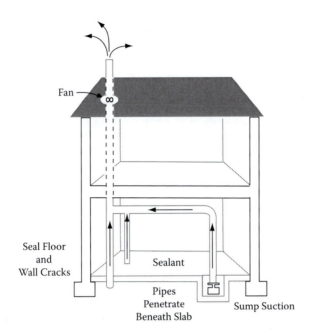

FIGURE 16.6
Components of a passive radon mitigation system. (Adapted from the U.S. Environmental Protection Agency, www.epa.gov/radon/pubs/rrnc-tri.html.)

intrusion mitigation, described in Chapter 7. The vertical pipes are then connected to the main riser that vents the vapors safely to the roof of the building. All unwanted pathways will be sealed to prevent uncontrolled entry of gases into the building.

References

Bell, William. 1992. Radon: What You Don't Know Can Hurt You! Massachusetts Department of Public Health.

Oregon Department of Health Services. Radon Protection Services. www.oregon. gov/DHS/ph/rps/radon/awarenessmonth.shtml.

Partington, J. R. 1957. Discovery of radon. *Nature,* 179(4566): 912.

U.S. Environmental Protection Agency. August 1986. A Citizen's Guide to Radon: What It Is and What to Do About It. OPA-86-004.

U.S. Environmental Protection Agency, Office of Air and Radiation. July 1992. Indoor Radon and Radon Decay Product Measurement Device Protocols.

U.S. Environmental Protection Agency. July 2000. Home Buyer's and Seller's Guide to Radon. 402-K-00-008.

U.S. Environmental Protection Agency. January 2009. A Citizen's Guide to Radon. http://www.epa.gov/radon/pubs/citguide.html.

U.S. Environmental Protection Agency. 2011. Decoy Chains. www.epa.gov/rpdweb 00/understand/chain.html.

17

Indoor Air Quality

17.1 Introduction

Indoor air quality is a broad topic that pulls in many topics discussed elsewhere in this book, such as asbestos, radon, lead (from deteriorated paint), and mold, as well as many of the chemical pollutants discussed in the book's early chapters. Because of the broadness of the issue, *indoor air quality* (IAQ) can be a very difficult problem to diagnose, even though the fixes are in many cases routine and uncomplicated.

Indoor air quality concerns generally occur in residential and office scenarios. IAQ issues may arise in industrial settings, but typically those hazards fall under the purview of the United States Occupational Safety and Health Administration (OSHA) and are not discussed in this chapter. The investigation and remediation of indoor air quality problems is quite interdisciplinary, bringing in toxicology as well as many engineering disciplines, especially mechanical engineering. This chapter focuses on the contributions that can be made by the environmental consultant to the identification and remediation of indoor air quality problems.

It should be noted that many of the contaminants that create IAQ problems are also regulated under the Clean Air Act. However, the Clean Air Act focuses on regional and area pollution and air pollution sources. It does not apply to indoor air problems, which are far more local in nature and impact.

17.1.1 History of Indoor Air Problems

Until the end of World War II, commercial buildings in the United States typically had windows that could be opened manually by the building occupants. Air conditioning, if available, was supplied by window-mounted units. This type of construction changed postwar, as commercial buildings increasingly were constructed with exterior walls that were permanently sealed. Central heating, ventilation, and air-conditioning (HVAC) systems became the norm. The trend toward fully sealed commercial buildings greatly accelerated in the 1970s, when the energy crisis increased the desire for energy efficient buildings, that is, buildings with minimum air leakage through rickety windows. Individuals lost direct control over their breathing air conditions, and the ubiquity of HVAC systems resulted in a "one air type fits all" for all building occupants.

The issue of indoor air quality reached national prominence in 1976 with the outbreak of what became known as *Legionnaire's disease* at an American Legion convention in Philadelphia, Pennsylvania. The *Legionellosis* bacterium was identified and found to be breeding in the cooling tower of the hotel's air conditioning system, from which it spread through the entire building. This event prompted a call for regulating climate control systems.

With the rise in indoor air quality complaints, by the early 1990s, the term *sick building syndrome* came into vogue. This term referred to buildings that were inherently defective, resulting in conditions ranging from uncomfortable to downright toxic to its inhabitants. Its appearance was enhanced by the increasing trend to use building materials that were chemically treated or that emitted chemicals when new, especially *volatile organic compounds* (VOCs), which are described in Chapter 3 and discussed in Section 17.2.5. Indoor air quality, as an environmental problem, was here to stay.

17.1.2 Health Effects of Indoor Air Pollutants

Health-related effects from an indoor air pollutant can result from a single exposure, multiple exposures, or prolonged, constant exposures. Symptoms typically include irritation to the eyes, nose, and throat, headaches, dizziness, and fatigue. Long-term exposure to indoor contaminants can lead to more serious illnesses, including respiratory diseases, heart disease, and cancer.

As with mold (see Chapter 16), most indoor air pollutants affect the most susceptible populations, including children; the elderly; and people suffering from asthma, allergies, and immunocompromised illnesses (as well as contact lens wearers). This is buttressed by the fact that many indoor contaminants such as mold lack an established dose-response relationship between exposure and health effects, which makes it difficult to connect the symptoms with the contaminant. Because of the amount of time the typical office worker spends inside the sealed modern office building (as opposed to the common residence), office workers are most likely to be affected by IAQ problems.

17.1.3 Indoor Air Investigation Triggers

IAQ investigations are ordinarily triggered by real or perceived health impacts or odor complaints by a resident or a building worker. IAQ investigations may also be triggered by a property transaction or financing, in which case they are performed as supplements to a Phase I Environmental Site Assessment, or as part of a risk assessment, often for a lending institution or insurance company.

17.2 Sources of Indoor Air Pollution

It is helpful to conceptualize indoor air pollutants as either originating inside the building or originating outside the building and somehow being drawn into the building. They all share inhalation as their primary pathway. Many indoor air pollutants, such as asbestos, radon, and mold, are discussed in other chapters in this book and are therefore not discussed in this chapter. The following discussions describe the other important categories of indoor air pollutant sources.

17.2.1 Poor Air Flow

Improperly maintained or improperly operated ventilation systems is a prime source of indoor air quality problems. It can be building wide, or, in the case of unbalanced ventilation systems, present in a portion or portions of a building. The air flow rates recommended by the American Society of Heating, Refrigerating and Air-Conditioning Engineers, Inc. (ASHRAE) are 15 cubic feet per meter (cfm) per person in an industrial setting and 20 cfm per person in an office setting. The generally accepted range for relative humidity is 30% to 50%, although it may vary based on the time of year (see later).

Poor air flow can cause a buildup of carbon dioxide (CO_2) from human exhalation. ASHRAE recommends that indoor CO_2 concentrations not exceed 700 parts per million (ppm). As of this writing, the generally accepted outdoor concentration of CO_2 at sea level in the United States is 385 ppm. This means that an indoor contribution of 1085 ppm, or 700 ppm above background, is needed for the indoor air to exceed the ASHRAE allowable CO_2 concentration of 700 ppm.

The range of indoor temperatures and relative humidity recommended by ASHRAE varies by the time of year and the type of indoor environment. However, even within those ranges, there will be people who complain about the temperature and humidity of the indoor air. For that reason, ASHRAE makes its recommendation using an 80% "occupant acceptability," meaning that they assume that 10% of all people will believe that the temperature or relative humidity are too low (10%) or too high (another 10%) at any given time, even with a properly operating ventilation system.

17.2.2 Combustion Products

Many indoor air pollutants are the byproducts of incomplete combustion of heating fuel. Sources include the building's heating system (if it burns fossil fuel), outdoor maintenance equipment, such as lawn mowers, snow and leaf blowers, and motor vehicles, especially in a residence with an attached garage. The byproducts of incomplete hazards include carbon monoxide

(CO), nitrogen oxide (NO), nitrogen dioxide (NO_2), and various volatile and semivolatile organic gases.

Carbon monoxide is a colorless, odorless gas that deprives the brain of oxygen, which can lead to nausea, unconsciousness and death. OSHA's one-hour exposure limit for CO in the workplace is 50 ppm, although the National Institute for Occupational Safety and Health (NIOSH) recommends a limit of 35 ppm. The American Conference of Governmental Industrial Hygienists (ACGIH) recommends an 8-hour time-weighted average (TWA) limit of 25 ppm for CO exposure.

Exposure to NO and NO_2 can irritate the respiratory tract, although high levels of exposure, unlikely in an indoor air setting, can cause more severe and permanent damage. There are no standards for NO and NO_2 in indoor air.

17.2.3 Dust and Particulates

Particulate emissions from incomplete combustion can cause respiratory irritation. Dust and particulates can contain contaminants, such as heavy metals (including lead) and asbestos, which can create their own set of problems. Particulates are usually described by their average diameter. For instance, PM_{10} *particles* are generally less than 10 microns in diameter. The USEPA recommended limit for PM_{10} in indoor air is 150 micrograms per cubic meter ($\mu g/m^3$) for a 24-hour TWA.

17.2.4 Ozone

Ozone is a critical, useful component of the earth's upper atmosphere. When down near the earth's surface, it is a pollutant rather than a desirable compound. Ozone can cause breathing problems because it is an irritant and because it is able to react with many chemicals found indoors, sometimes creating chemicals that are toxic to humans. Organic compounds formed by the reactions between ozone and many common indoor pollutants including organic compounds can be more odorous, irritating, or toxic than the pollutants they derive from. Recent research has shown that indoor ozone levels in the breathable air should not exceed 20 parts per billion (ppb).

17.2.5 Volatile Organic Compounds

VOCs are emitted from literally thousands of solid or liquid products and byproducts. They may have short- and long-term adverse health effects, depending on the chemical.

VOCs are widely used as ingredients in household products, such as paints and lacquers, paint strippers, cleaning supplies, waxes, mothballs, and air fresheners. Chemicals typically found in garages and maintenance areas, such as gasoline and other fuels, automotive products, and industrial cleaners, can contain VOCs. VOCs are in office products, such as inks, correction fluids, and office equipment such as copiers and printers. They also can be combustion byproducts or byproducts of chemical reactions.

Another source of VOCs common to both offices and residences is dry-cleaned clothing. Most dry cleaners use perchloroethylene (also known as PCE or perc) to clean clothing. Not all of the perc is removed from the clothing in the drying process, so dry-cleaned clothing in a closet or on a person is a potential source of VOCs. This compound is also important to site investigation (see Chapter 6) and vapor intrusion studies (see Chapter 7).

VOCs may also originate as off-gasses from building materials. Formaldehyde can off-gas from carpeting (most carpets now on the market are advertised as "low VOC"); resins from pressed wood products, whose usage has greatly increased in recent years; *urea formaldehyde foam insulation* (UFFI), which was used as insulation in many homes in the 1970s; and paints and varnishes.

17.2.6 Bioaerosols

Biological hazards typically are *bioaerosols*, which are biological agents that remain suspended in air. They include mold (discussed in Chapter 15), viruses, pollen, dust mites, and animal hair and dander, all of which can produce a variety of health-related effects in susceptible populations. *Legionella* is a bioaerosol of special concern, as mentioned at the beginning of the chapter. It grows in slow-moving or still, warm water, such as water found in evaporative cooling towers or showerheads, thus making it a hazard to commercial as well as noncommercial buildings.

17.2.7 Tobacco Smoke

Environmental tobacco smoke (ETS) is caused by smoking inside a building or smoking near the ventilation intake of a building. It is becoming rarer now since smoking bans inside buildings are becoming more common. In large quantities, second-hand smoke can be as cancer-causing to the bystander as the smoker. From an IAQ perspective, it can cause respiratory irritation and can react with ambient ozone to form deadly carbon monoxide.

17.2.8 Pesticides

Pesticides, either from application inside the building, storage inside the building, or migrating inside the building from an outdoor application, can affect indoor quality. Of particular concern is the indoor application of pesticides meant to be used outdoors, since they are likely to persist indoors and be a constant threat to human health.

17.2.9 Subsurface Contamination

Vapor intrusion, an indoor air quality issue that is triggered by vapors emanating from soil or groundwater that is contaminated with chemicals, usually VOCs, is discussed in Chapter 7.

17.3 Heating, Ventilation, and Air-Conditioning (HVAC) Systems

Since the building's HVAC system is such a crucial component to the whole indoor air issue, it pays to have a working understanding of how the system works in order to perform a proper IAQ study.

The HVAC system includes all heating, cooling, and ventilation equipment serving a building. The heating and cooling systems generally run independent of each other, although both systems, as well as the ventilation system, use a central component known as the *air handler* (see Figure 17.1). The air handler brings in outdoor air, mixes it with air that has circulated through the building, heats or cools the air as needed, sends it into the building, and then retrieves it for reconditioning and reuse.

Outdoor air enters the air handler through the air intake, which will allow in a minimum amount of outdoor air during extreme outside temperature conditions. Air intakes are equipped with air filters, which come in a variety of efficiencies, but generally are designed to keep out only particulates and not gases. Outdoor air and recirculated air are mixed in the air handler, and then sent through heating and cooling coils to regulate the temperature of the air delivered to the space.

Figure 17.2 is a schematic diagram of the air-conditioning ("chiller") portion of an HVAC system. A compressor compresses the air, thus heating it. A condenser removes moisture from the air, after which it goes through an expansion valve, which cools the air. The cooled air is then blown into the air distribution system by a supply fan. Air is usually distributed in commercial buildings through ducts located above suspended ceilings and behind walls. The air circulates through the user space, and then is pulled through a ceiling-mounted return into ducts and returned back to the air handler (see Figure 17.3). In some cases, the space above the dropped ceiling itself is used for returning air to the air handler. This is done as a cost-saving device, but can cause materials or growths above the dropped ceiling to mix into the distributed air and spread throughout the building.

Two portions of the air-conditioning system can spawn bacterial growth if not properly maintained. The condensate created by the condenser, as described earlier, collects in the drain pan under the cooling coil and exits the air handler via a deep seal trap. If not allowed to dry periodically or allowed to build up cellulosic materials they can be a breeding ground for

FIGURE 17.1
A typical industrial air-handling unit.

bioaerosols. The same holds true for the cooling tower, which is part of a commercial air-conditioning system (see Figure 17.2).

Some buildings are kept under slightly positive air pressure relative to the outdoors to reduce infiltration of outside air, which could result in moisture and high humidity inside the building. For that reason, air flow from the air handler is sometimes greater than the air flow back to the air handler.

17.4 Performing the Indoor Air Quality Investigation

More than any other type of investigation discussed in this book, the design of IAQ investigation depends greatly on the type of air quality problem.

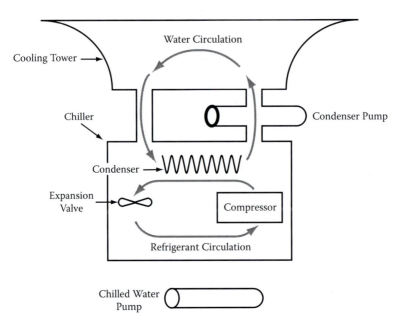

FIGURE 17.2
A schematic diagram of a commercial air-conditioning system.

Therefore, it is imperative that the investigator gain a thorough knowledge of the nature of the IAQ problem and be able to place the problem in a physical and temporal context. Otherwise, the investigation is doomed to failure.

Investigations typically fall into one of three categories: epidemiological, environmental, or engineering. An *epidemiological IAQ assessment* will emphasize the health-related aspects of the triggering event and will seek the cause or causes of the outbreak. An *environmental IAQ assessment* will look for the source of the odor complaints or some other triggering event relating to the storage or usage of chemicals in and around the building. The *engineering IAQ assessment* will look for air flow problems and seek their remedy. The methods used to perform the steps discussed next will depend on the given situation and which category of investigation is being conducted.

17.4.1 Building Inspection

An initial walk-through of the problem area provides information about the four basic factors influencing indoor air quality: occupants, contaminant sources, pollutant pathways, and the air circulatory system. It may be helpful to schedule the walk-through to coincide with the time of the day that the complaint occurred or conduct multiple walk-throughs in the course of the day if you suspect that timing is of issue.

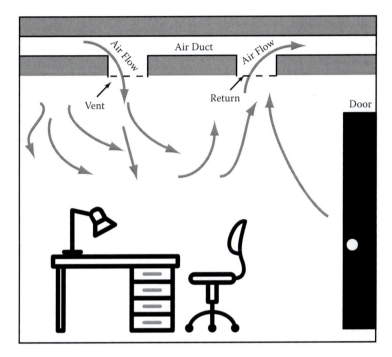

FIGURE 17.3
A supply and return air system in an office.

Initial observations are often the best. The inspector should assess quality of air in the building and the presence of objectionable odors. Is the air stale, too dry, or too damp? Is it too hot or too cold inside the building, or are there large temperature variations depending on where you are inside the building, or the time of day? Is outdoor air entering the building through uncontrolled pathways, such as leaking windows, doors, and gaps in the exterior construction?

The quality of the breathable area should be noted, particularly in the area of the complaint. The air circulation system itself could be the source of the problem, and should be studied in detail, especially as a potential pathway for contaminants, if the problem is widespread. Investigating the air distribution system should begin with the components of the HVAC system that serves the complaint area and surrounding rooms but may need to expand if connections to other areas are discovered. Is there evidence of deterioration, corrosion, water damage, or excessive dust or debris within the system? Does the air distribution system suffer from blocked vents, inoperative fans, or clogged filters?

Where health-related issues are involved, the inspector should check the drain pans for a buildup of water that can allow bacteria and mold to breed. The inspector should check around the air intakes for pollution sources, since rooftop or wall-mounted air intakes can pose a problem if they are located near or downwind of exhaust outlets or other contaminant sources.

For most commercial buildings, there is ample available information on the HVAC system's construction and its maintenance. The investigator should review available construction and operating records, and collect detailed information on the HVAC system and possible pollutant pathways. Material safety data (MSD) sheets, which are required to be readily available at the facility, provide information regarding the chemicals used by the building maintenance staff in the HVAC system and other building equipment.

Understanding the building occupants and related contaminant sources should help to identify potential sources of odors/chemicals, be it chemicals used by employees, chemicals worn by employees, or materials that are new and might be emitting off-gasses. It is important to observe office equipment, particularly large office equipment such as engineering drawing reproduction machines and wet-process copiers, learn about their usage patterns, and identify items that are not equipped with local exhaust.

The combustion equipment used by the building would be a prime suspect if combustion products are the contaminants of concern. Areas with water damage that could result in mold growth should be identified, as well as the locations of new furnishings that could be emitting VOCs or other gases. The inspector should evaluate the quality of housekeeping practices, and look for deteriorated or moldy building components and equipment.

The building vicinity must also be inspected for contaminant sources if it is possible that chemicals are entering the building through its air intake. The investigator should observe patterns of traffic, construction activity, and other potential sources in the neighborhood of the building, and inquire about outdoor ambient air problems in the area.

17.4.2 Interviews

A critical piece of the IAQ assessment is the interview phase. There is no substitute for getting the critical information directly from the source. The interviewer must confirm the health-related issue, which could range from one mildly ill person to a building-wide epidemic, by interviewing the people who have lodged the complaints. The symptoms should be described as precisely as possible, then compared with the medical histories of the affected persons, so correlations can be made.

Since the information provided by those interviewed can be biased and conflicting, the interviewer should be skeptical about the information gathered during the interview. In many cases, people believe that they are suffering from indoor air pollutants, when in fact they are suffering from some other problem not related to the indoor air. Checking whether other people in the work area are suffering from similar symptoms may help the interviewer connect the dots. Because workplace stresses can manifest themselves as health complaints, the interviewer must be aware of the context of the complaints during interviews.

Speaking with building personnel will help the interviewer to obtain a history of the building as it relates to indoor air issues. Personnel could be asked if there have been any recent fires, roof leaks, or smoke damage inside the building. Recent modifications, maintenance, or problems with the building's HVAC system are important to know.

If it appears that the problem could be due to elevated carbon monoxide levels or the presence of *Legionella*, the situation should be treated as an emergency and the investigator should leave the premises along with the other building occupants.

17.4.3 Making a Diagnosis

When diagnosing the indoor air problem, it is important to consider the location and timing of the complaints. The investigator must look for patterns in the timing of the complaints, in the symptoms, and in the location of the complaints. Did the onset of the symptoms coincide with an indoor pesticide application? Are they cyclic, for instance, are they present in the morning then gradually taper off in the afternoon? Has there been a change in the room use or the number of occupants in the room? Has the room been renovated, say, with the installation of new carpeting?

If the complaints are widespread, with no spatial pattern, then the building's HVAC system is a prime suspect. It is also possible that contaminated outdoor air is finding its way inside the building, or some objectionable cleaning material is being used in the building. Localized problems tend to have a localized source: some chemical in the area or maybe a problem with the HVAC's air vent or return.

If the complaints lack consistency, that is, spatially related but with different symptoms, the investigator must ask the following: Could these diverse symptoms have the same cause? Are the symptoms continuous? Are they present outside the workplace? Do they start when the individuals show up to work, or are they still present when the individuals have been away from the office for an extended period of time?

If a single individual has the problem, then the problem may have nothing to do with the building. If a localized problem cannot be identified, then the individual may be mistaking a personal problem for a problem related to indoor air quality.

17.5 Air Measurement Methods

Air measurements are usually performed as part of an engineering IAQ assessment. If poor air flow is initially suspected as the cause of the IAQ problems, the investigator can go right to measuring the air and skipping most of the aforementioned steps. In fact, when poor air flow is suspected, air measurements are often collected on the initial building walk-through.

17.5.1 Air Measurement Devices

Air is generally measured using direct reading instruments, that is, instruments specifically designed for collecting one type or a few limited types of data. For instance, carbon dioxide is measured using a CO_2 monitor. The direct reading instrument shown in Figure 17.4 can measure CO_2, temperature, humidity, and carbon monoxide simultaneously, making it and meters like it useful tools when collecting air measurements.

All quantitative measuring instruments should be calibrated in accordance with manufacturers' specifications before they are used. The instrument calibration should be documented in the project files.

Chemical smoke tubes can be used to determine pollutant pathways in the complaint area, between sections of the building, and outside the building. A micromanometer (or equivalent) can measure the magnitude of pressure differences between these areas. Switching air handlers or exhaust fans on and off, opening and closing doors, and simulating the range of operating conditions in other ways can help to show the different ways that airborne contaminants move within the building.

Scented oils, whose smell can be tracked throughout the study area, are sometimes used to determine pollutant pathways. However, the nose can quickly become desensitized to odors, making this method unreliable. Using

FIGURE 17.4
Collecting multiple gas readings using a direct reading instrument.

a tracer gas such as sulfur hexafluoride (SF_6) can provide qualitative as well as quantitative information on air flow patterns inside a building.

17.5.2 Locations of Air Measurements

The investigator should collect indoor air measurements in portions of the building where the occupants are known to spend their time and that are most representative of occupant conditions in that part of the building. If representative areas cannot be identified, then measurements should be collected in the middle of the room or zone. If temperature is the issue, measurements should be collected in locations where the most extreme values of temperature are likely to occur, such as near doors and windows, and corners away from and close to heating or cooling equipment. Collecting temperature measurements from multiple heights above the floor tests for vertical asymmetry. Absolute humidity only needs to be measured at one location in a building with an HVAC system, since humidity distribution usually evens out in a short period of time.

17.6 Air Sampling Methods

The decision to collect air samples for laboratory analysis should be made with caution. Often, because there are a large number of potential sources and a variety of contaminants present in the ambient air inside a building, air sampling results can raise more questions than they answer. Before collecting air samples, the investigator should have a clear understanding of what substances should be tested for and have given thought to how the results will be interpreted prior to running the tests. This thought process will help the investigator select appropriate parameters to test for, and appropriate locations and times to collect the air samples.

17.6.1 Air Sampling Locations

Once the chemicals of interest have been selected through the various means discussed earlier in this chapter, the investigator must decide where and when to collect the samples. Samples may be biased toward "worst-case" conditions, such as measurements during periods of maximum equipment emissions, minimum ventilation, or disturbance of contaminated surfaces. Worst-case sample results can be very helpful in characterizing maximum concentrations to which occupants are exposed and identifying sources for corrective measures.

Often it is advisable to obtain samples during average or typical conditions as a basis of comparison. It may, however, be difficult to know what conditions are typical. Research shows that exposure to some pollutants may vary dramatically as building conditions change. Devices that allow continuous measurements of key variables can be helpful.

17.6.2 Air Sampling Devices

The most frequently used method for testing for a variety of pollutants involves the use of *summa canisters*. Summa canisters, which are also discussed in the vapor intrusion section of Chapter 7, are stainless steel chambers that are quite versatile in their usage in indoor air investigations. Devoid of air when they arrive at a site, they are fitted with a regulator that can be adjusted to allow air flow into the canister. When the canister is full, the regulator is closed and the canister is shipped to a laboratory for analysis. The air inside the canister can be analyzed for VOCs, or for pesticides, PCBs, polycyclic aromatic hydrocarbons (PAHs), dioxin, and many other classes of chemicals using other USEPA-approved analytical methods.

Testing for *Legionella* typically involves the collection of water samples or surface samples when the bacterium's presence is suspected, which are generally places where warm water is present. Samples are usually incubated and analyzed in the same manner as mold samples (see Chapter 15).

One popular method for testing for a specific compound is the use of Draeger tubes. Draeger (also spelled Dräger) tubes are glass tubes that are filled with a chemical that will react with the compound of interest. They are connected to a pump, which is designed to draw air into the attached tube at a set pumping rate. As the reactant is exposed to the compound of interest, it changes color. The tube is graduated, so that the quantity of the reactant that changes color can be correlated with the concentration of the compound of interest in the air sample. Short-term tubes are designed for tests lasting 10 seconds to 15 minutes. The less common long-term tubes are designed to provide TWAs over the course of 1 to 8 hours.

17.7 Indoor Air Mitigation

Once the indoor air problem has been identified, its mitigation is usually pretty straightforward. Mitigation methods generally involve:

- Improved ventilation
- Air cleaning
- Pathway constriction or elimination
- Source reduction or elimination

Improving ventilation is often simply a matter of adjusting the air flow rate within the building, changing the amount of outdoor air entering the building, or removing obstructions to the air flow within the building. It may involve fixing broken or malfunctioning equipment or repairing leaks and seepage in the system. A licensed HVAC engineer should perform these activities; an untrained person modifying a building system can unintentionally create other problems.

Better air cleaning could be achieved by replacing or fixing broken or damaged equipment. Particulate control could be enhanced by installing electrostatic precipitators and particulate filters.

Removing the sources of the indoor air problems is often the simplest way of solving an indoor air quality problem. Improved housekeeping procedures can solve a myriad of indoor air quality problems. Vacuuming carpeting and fabric-covered furniture regularly can remove animal dander, dust mites, and other annoying and allergenic particulates from the workspace or residence. Removing dust from hard surfaces, toys, and so forth is also recommended. Pollutant reduction practices include improved chemical storage and handling procedures, changing to less toxic chemicals, and workplace rules limiting chemicals in personal care products, such as hairsprays. Techniques for controlling the pathways for pollutants may include keeping motor vehicles away from air intakes, and making sure that building heating systems that use fossil fuels are properly adjusted to ensure more complete combustion and adequate ventilation to the outdoors.

Remediation for *Legionella* outbreaks in commercial buildings often include flushing the source with hot water (>160°F; 70°C), sterilizing standing water in evaporative cooling basins, replacing shower heads, and in some cases flushing the source with heavy metal salts, which will kill the *Legionella*.

References

American Society of Heating, Refrigerating and Air-Conditioning Engineers, Inc. 2004. ASHRAE Standard for Thermal Environmental Conditions for Human Occupancy.

ASTM International. 2010. Standard Guide for Vapor Encroachment Screening on Property Involved in Real Estate Transactions. E-2600-10.

Bower, John. 2001. *Healthy House Building*. The Healthy House Institute.

Dräger Safety AG & Co. KGaA. 2008. Dräger-Tubes & CMS Handbook, 15th ed.

Interstate Technology & Regulatory Council. January 2007. Vapor Intrusion Pathway: A Practical Guideline.

New Jersey Department of Environmental Protection. August 2005. Vapor Intrusion Guidance.

Salthammer, Tunga, ed. 1999. *Organic Indoor Air Pollutants: Occurrence, Measurement, Evaluation*. Wiley-VCH.

Spengler, Jonathan D., and Samet, John M. 1991. *Indoor Air Pollution: A Health Perspective*. Baltimore, MD: Johns Hopkins University Press.

Spengler, Jonathan D., Samet, John M., and McCarthy, John F. 2001. *Indoor Air Quality Handbook*. New York: McGraw–Hill.

Tichenor, Bruce. 1996. Characterizing Sources of Indoor Air Pollution and Related Sink Effects. ASTM STP 1287.

U.S. Environmental Protection Agency and U.S. Consumer Product Safety Commission. April 1995. The Inside Story: A Guide to Indoor Air Quality. EPA Document #402-K-93-007.

U.S. Environmental Protection Agency et al. January 2005. IAQ Tools for Schools, 3rd ed. EPA Document #402-K-95-001.

U.S. Environmental Protection Agency, Office of Solid Waste and Emergency Response. November 2002. Draft OSWER Guidance for Evaluating the Vapor Intrusion to Indoor Air Pathway from Groundwater and Soils (Subsurface Vapor Intrusion Guidance). EPA-530-D-02-004.

U.S. Environmental Protection Agency, Office of Solid Waste and Emergency Response. 2010. Review of the Draft 2002 Subsurface Vapor Intrusion Guidance.

18

The Environmental Project

This final chapter of the book briefly describes the business side of the environmental consulting project.

18.1 Projects: It All Begins with a Contract

Since consultants, by definition, provide professional services to clients, there must be a contractual relationship between the consultant and client for work to proceed. It is useful to think of a standard consulting contract as having three parts: a scope of work, a schedule, and a budget. The *scope of work* defines the tasks that the consultant will provide to the client. The *schedule* includes interim as well as final benchmarks in the completion of the scope of work. The *budget* describes the compensation that the consultant will receive for its services. All three pieces are interrelated, as described next.

18.1.1 Types of Contracts

Contracts are awarded on a variable budget or a fixed-base budget. Costs fall into three categories:

- *Labor costs*, which result from time spent by the environmental consultant in executing the scope of work.
- *Pass-through costs*, which result from the consultant's use of outside services (vendors and subcontractors, as described in Section 18.1.2).
- *Other direct expenses*, which include travel costs, lodging and meals costs, and other incidental costs incurred by the consultant while performing the scope of work.

In *time-and-materials* (T&M) projects, both the client and the consultant agree in writing that the project costs may vary up or down in the course of the fulfillment of the contract. This type of contract is commonly employed when the scope of work and work conditions are unknown or likely to change. In fixed-base, or *lump-sum* projects, the budget that the consultant will charge the client is set, regardless of what happens while completing the agreed-upon scope of work. If the scope of work changes, or work conditions change in a manner that the client acknowledges is unforeseen and therefore deserving of additional compensation, the budget in a lump-sum contract is amended through the issuance of a *change order*.

Contracts that lack defined scopes of work and contracts with an undefined scope of work are known as *indefinite delivery/indefinite quantity* (IDIQ) contracts. These contracts are typically issued by public agencies; they do not commit the issuing agency to a particular scope of work or a particular expenditure of money, but establish the terms and cost structure by which future expenditures under the contract will be governed. IDIQ contracts typically expire after a defined period, after which the consultant would have to submit a new bid to the client for renewal of the contract.

18.1.2 The Project Team

The environmental consulting project typically is a team effort, involving many people at various levels within the organization. The larger the project, the greater the involvement in terms of number of consultants, their levels within the organization, and the skill sets they bring to the table. At the helm is a senior-level person who directs the project, and usually has the title of principal or equivalent within the organization. That person is responsible for putting together the project team and seeing that the team fulfills its contractual goals. The point person on the project team manages the scope of work, schedule, and project budget, and reports to the person directing the project. That person is assisted by various team personnel, who may be junior to and less experienced than the project manager, or may be technical experts who perform various defined tasks within their areas of expertise on the project.

Execution of a project typically involves both fieldwork (data gathering) and office work (data interpretation and report preparation). Performance of the fieldwork portion of the scope of work is often aided by outside parties known as contractors (sometimes called subcontractors) and vendors. *Contractors* generally perform fieldwork that the environmental consulting firm does not routinely do on its own. This work may involve the usage of heavy equipment, such as earth moving equipment and drilling equipment, or the performance of a specific technical task outside the field of expertise of the environmental consultant, such as asbestos or lead-based paint removal, or the installation of a vapor mitigation system. *Vendors* either provide project services at locations other than the project site, such as a fixed-base laboratory, or provide equipment to the project, either through purchase or rental. As with the client–environmental consultant relationship, contracts, usually known as *subcontracts*, govern the activities performed by subcontractors and vendors on a project.

18.1.3 Project Completion

When the agreed-upon scope of work has been completed, the consultant generally prepares a written document, often referred to as a *deliverable*. This document, which may be in the form of a report or a letter, indicates that the scope of work has been completed. It documents the methodologies and findings of the scope of work, and usually includes recommendations, if any, for follow-up activities.

The other document (electronic or otherwise) that is prepared at the end of the project is an *invoice*. This invoice may cover all or part of the project budget. In a fixed-base contract, the invoice will be expressed either as a set dollar figure or as a percentage of the agreed-upon budget. For time-and-materials contracts, the invoice will be based on actual labor hours spent, usually indicated as the hours-per-billing category (principal, project manager, etc.) multiplied by the agreed-upon billing rate for that labor category, plus contractor costs and expenses, which often are multiplied by a *mark-up factor*, which covers some of the administrative costs associated with the handling of so-called pass-through costs, that is money that is not retained by the consulting firm upon receipt, but rather passed on to the contractor or vendor who rendered the service to the environmental consultant.

Appendix A

List of Abbreviations

ACBM	asbestos-containing building material
ACM	asbestos-containing material
AE	acid extractable compound
AHERA	Asbestos Hazard Emergency Response Act
AOC	area of concern
APR	air-purifying respirator
ARARs	applicable or relevant and appropriate requirements
AS	air sparging
ASHARA	Asbestos School Hazard Abatement Reauthorization Act
ASHRAE	American Society of Heating, Refrigerating and Air-Conditioning Engineers, Inc.
AST	aboveground storage tank
ASTM	American Society for Testing & Materials (formerly)
AUL	activity and use limitation
BAF	bioaccumulative factor
BAT	best available technology
BN	base-neutral compounds
BOD	biochemical oxygen demand
BSAF	sediment/soil-to-biota bioaccumulation factor
BTEX	benzene, toluene, ethylbenzene, and xylenes
CAA	Clean Air Act
CATEX	categorical exclusion
CERCLA	Comprehensive Environmental Response, Compensation, and Liability Act
CEU	continuing education unit
CFR	Code of Federal Regulations
CFS	cubic feet per second
CFU	colony forming units
CHMM	Certified Hazardous Materials Manager
CIH	Certified Industrial Hygienist
COD	chemical oxygen demand
CORRACTS	RCRA Corrective Action database
CWA	Clean Water Act
dBA	A-weighted noise decibel
DEIS	Draft Environmental Impact Statement
DNAPL	dense nonaqueous phase liquid
DO	dissolved oxygen

EA	environmental assessment
ECSM	ecological conceptual site model
EIS	environmental impact statement
EP	Environmental Professional
EPCRA	Emergency Planning and Community Right-to-Know Act
ESA	Endangered Species Act
ESA	environmental site assessment
FEIS	final environmental impact statement
FIFRA	Federal Insecticide, Fungicide, and Rodenticide Act
FOIA	Freedom of Information Act
FONSI	finding of no significant impact
FS	feasibility study
GAC	granular activated carbon
GIS	geographic information system
GPR	ground penetrating radar
HAPs	hazardous air pollutants
HASP	health and safety plan
HAZWOPR	Hazardous Waste Operations
HEPA	high-efficiency particulate air
HMI	hazardous materials inventory
HREC	historical recognized environmental condition
HRS	Hazard Ranking System
HUD	U.S. Department of Housing and Urban Development
HVAC	heating, ventilation, and air-conditioning
IAQ	indoor air quality
IDIQ	indefinite delivery/indefinite quantity
IRIS	Integrated Risk Information System
ISCO	*in situ* chemical oxidation
ISO	International Standards Organization
LBP	lead-based paint
LEDPA	least environmentally damaging practicable alternative
LEED	Leadership in Energy and Environmental Design
LF	linear feet
LNAPL	light nonaqueous phase liquid
LQG	large quantity generator
LUST	leaking underground storage tank
MCL	maximum contaminant level
MCLG	maximum contaminant level guideline
MNA	monitored natural attenuation
MSD sheet	material safety data sheet
MSL	mean sea level
MSW	municipal solid waste
NAAQS	National Ambient Air Quality Standards
NAPL	nonaqueous phase liquid
NCP	National Contingency Plan

NEPA	National Environmental Protection Act
NESHAPS	National Emissions Standards for Hazardous Air Pollutants
NIOSH	National Institute for Occupational Safety and Health
NOB	nonorganically bound
NPDES	National Pollution Discharge Elimination System
NPL	National Priorities List
NPS	nonpoint sources
NSDWR	National Secondary Drinking Water Regulations
NTU	nephelometric turbidity units
O&M	operations and maintenance
ORP	oxygen reduction potential
OSHA	U.S. Occupational Safety and Health Administration
PACM	presumed asbestos-containing materials
PAH (also PNA)	polycyclic aromatic hydrocarbons (also known as polynuclear aromatics)
PCBs	polychlorinated biphenyls
PCM	phase contrast microscopy
PE	Professional Engineer
PEL	permissible exposure limit
PG	Professional Geologist
PID	photoionization detector
PLM	polarized light microscopy
POET	point of entry treatment
POU	point of usage
PPE	personal protective equipment
PRP	potentially responsible party
PSD	prevention of significant deterioration
PVC	polyvinyl chloride
QA	quality assurance
QC	quality control
RCRA	Resource Conservation and Recovery Act
REC	recognized environmental condition
RfC	reference concentration
RfD	reference dose
RI	remedial investigation
ROD	record of decision
RQD	rock quality designation
SAP	sampling and analysis plan
SAR	supplied-air respirator
SARA	Superfund Amendments and Reauthorization Act
SDWA	Safe Drinking Water Act
SEIS	Supplemental Environmental Impact Statement
SF	square feet
SHWS	state hazardous waste site
SITE	Superfund Innovative Technology Evaluation
SOCs	synthetic organic compounds

SQG	small quantity generator
SVE	soil vapor extraction
SVOCs	semivolatile organic compounds
SWL	solid waste landfill
T&M	time and materials
TAL	Target Analyte List
TCL	Target Compound List
TCLP	toxicity characteristic leachate procedure
TEM	transmission electron microscopy
TICs	tentatively identified compounds
TMDL	total maximum daily load
TOC	total organic carbon
TPH	total petroleum hydrocarbons
TPH-DRO	total petroleum hydrocarbons–diesel range organics
TPH-GRO	total petroleum hydrocarbons–gasoline range organics
TRI	toxic release inventory
TRV	toxicity reference values
TSCA	Toxic Substances Control Act
TSDF	treatment, storage, and disposal facility
TSI	thermal system insulation
TWA	time-weighted average
USACE	U.S. Army Corps of Engineers
USEPA	U.S. Environmental Protection Agency
USGS	U.S. Geological Survey
UST	underground storage tank
VOC	volatile organic compound
XRF	x-ray fluorescence

Appendix B

State Environmental Departments

State	State Environmental Agency	Acronym	Web Address
Alabama	Department of Environmental Management	ADEM	www.adem.state.al.us
Alaska	Department of Environmental Conservation	ADEC	www.dec.state.ak.us/
Arizona	Department of Environmental Quality	ADEQ	www.azdeq.gov/
Arkansas	Department of Environmental Quality	ADEQ	www.adeq.state.ar.us/
California	California Environmental Protection Agency	Cal EPA	www.calepa.ca.gov/
Colorado	Department of Public Health and Environment	CDPHE	www.cdphe.state.co.us/
Connecticut	Department of Environmental Protection	DEP	www.ct.gov/dep/site/default.asp
Delaware	Department of Natural Resources and Environmental Control	DNREC	www.dnrec.delaware.gov
Florida	Department of Environmental Protection	DEP	www.dep.state.fl.us/
Georgia	Department of Natural Resources	GADNR	www.gadnr.org/
Hawaii	Department of Health	HDOH	www.hawaii.gov/health/
Idaho	Department of Environmental Quality	DEQ	www.deq.idaho.gov/
Illinois	Illinois Environmental Protection Agency	IEPA	www.epa.state.il.us/
Indiana	Department of Environmental Management	IDEM	www.in.gov/idem/
Iowa	Department of Natural Resources	DNR	www.iowadnr.gov/
Kansas	Department of Health and Environment	KDHE	www.kdheks.gov/environment

State	State Environmental Agency	Acronym	Web Address
Kentucky	Department for Environmental Protection	DEP	www.dep.ky.gov
Louisiana	Department of Environmental Quality	DEQ	www.deq.louisiana.gov/portal/
Maine	Department for Environmental Protection	DEP	www.maine.gov/dep
Maryland	Maryland Department of the Environment	MDE	www.mde.state.md.us
Massachusetts	Massachusetts Department of Environmental Protection	MassDEP	www.mass.gov/dep/
Michigan	Department of Environmental Quality	DEQ	www.michigan.gov/deq
Minnesota	Minnesota Pollution Control Agency	MPCA	www.pca.state.mn.us/
Mississippi	Department of Environmental Quality	MDEQ	www.deq.state.ms.us/
Missouri	Department of Natural Resources	DNR	www.dnr.mo.gov/env/index.html
Montana	Department of Environmental Quality	DEQ	www.deq.mt.gov/default.mcpx
Nebraska	Nebraska Department of Environmental Quality	NDEQ	www.deq.state.ne.us/
Nevada	Nevada Division of Environmental Protection	NDEP	www.ndep.nv.gov/
New Hampshire	Department of Environmental Services	DES	www.des.nh.gov/
New Jersey	Department of Environmental Protection	NJDEP	www.state.nj.us/dep/
New Mexico	New Mexico Environmental Department	NMED	www.nmenv.state.nm.us/
New York	Department of Environmental Conservation	NYSDEC	www.dec.ny.gov/
North Carolina	Department of Environment and Natural Resources	NCDENR	http://portal.ncdenr.org/web/guest
North Dakota	Department of Health—Environmental Division	DoH	www.ndhealth.gov/EHS/
Ohio	Ohio Environmental Protection Agency	Ohio EPA	www.epa.state.oh.us/

State	State Environmental Agency	Acronym	Web Address
Oklahoma	Department of Environmental Quality	DEQ	www.deq.state.ok.us/
Oregon	Department of Environmental Quality	DEQ	www.oregon.gov/DEQ/
Pennsylvania	Department of Environmental Protection	DEP	www.depweb.state.pa.us/
Rhode Island	Department of Environmental Management	DEM	www.dem.ri.gov/
South Carolina	Department of Health and Environmental Control	DHEC	www.scdhec.gov/
South Dakota	Department of Environment and Natural Resources	DENR	www.denr.sd.gov/
Tennessee	Department of Environment and Conservation	TDEC	www.state.tn.us/ environment/
Texas	Texas Commission on Environmental Quality	TCEQ	www.tceq.state.tx.us/
Utah	Department of Environmental Quality	DEQ	www.deq.utah.gov/
Vermont	Department of Environmental Conservation	DEC	www.anr.state.vt.us/dec/dec. htm
Virginia	Department of Environmental Quality	DEQ	www.deq.state.va.us/
Washington	Washington State Department of Ecology	WA DOE	www.ecy.wa.gov/ecyhome. html
West Virginia	Department of Environmental Protection	DEP	www.dep.wv.gov/Pages/ default.aspx
Wisconsin	Department of Natural Resources	WDNR	www.dnr.wi.gov/index.asp
Wyoming	Department of Environmental Quality	DEQ	http://deq.state.wy.us/
Washington, DC	District Department of the Environment	DDOE	www.ddoe.dc.gov/ddoe/site/

Index